Emmanuel Sobgwi Tamne

**The Reactivity Studies of Chromium-Chromium Quintuple Bond**

Emmanuel Sobgwi Tamne

# The Reactivity Studies of Chromium-Chromium Quintuple Bond

## Stabilized by Aminopyridinato Ligands

Südwestdeutscher Verlag für Hochschulschriften

**Impressum / Imprint**
Bibliografische Information der Deutschen Nationalbibliothek: Die Deutsche Nationalbibliothek verzeichnet diese Publikation in der Deutschen Nationalbibliografie; detaillierte bibliografische Daten sind im Internet über http://dnb.d-nb.de abrufbar.
Alle in diesem Buch genannten Marken und Produktnamen unterliegen warenzeichen-, marken- oder patentrechtlichem Schutz bzw. sind Warenzeichen oder eingetragene Warenzeichen der jeweiligen Inhaber. Die Wiedergabe von Marken, Produktnamen, Gebrauchsnamen, Handelsnamen, Warenbezeichnungen u.s.w. in diesem Werk berechtigt auch ohne besondere Kennzeichnung nicht zu der Annahme, dass solche Namen im Sinne der Warenzeichen- und Markenschutzgesetzgebung als frei zu betrachten wären und daher von jedermann benutzt werden dürften.

Bibliographic information published by the Deutsche Nationalbibliothek: The Deutsche Nationalbibliothek lists this publication in the Deutsche Nationalbibliografie; detailed bibliographic data are available in the Internet at http://dnb.d-nb.de.
Any brand names and product names mentioned in this book are subject to trademark, brand or patent protection and are trademarks or registered trademarks of their respective holders. The use of brand names, product names, common names, trade names, product descriptions etc. even without a particular marking in this works is in no way to be construed to mean that such names may be regarded as unrestricted in respect of trademark and brand protection legislation and could thus be used by anyone.

Coverbild / Cover image: www.ingimage.com

Verlag / Publisher:
Südwestdeutscher Verlag für Hochschulschriften
ist ein Imprint der / is a trademark of
OmniScriptum GmbH & Co. KG
Heinrich-Böcking-Str. 6-8, 66121 Saarbrücken, Deutschland / Germany
Email: info@svh-verlag.de

Herstellung: siehe letzte Seite /
Printed at: see last page
**ISBN: 978-3-8381-3923-4**

Zugl. / Approved by: Bayreuth, Uni-Bayreuth, Diss., 2013

Copyright © 2014 OmniScriptum GmbH & Co. KG
Alle Rechte vorbehalten. / All rights reserved. Saarbrücken 2014

*In memory of my mother*

| | |
|---|---|
| Ar | aryl |
| Ap | aminopyridine |
| Å | Angstrom |
| BuLi | butyl lithium |
| br | broad |
| °C | degree Celsius |
| ° | degree |
| d | doublet |
| δ | chemical shift (ppm) |
| Et | ethyl |
| g | gram |
| h | hours |
| Hz | Hertz |
| $J$ | Coupling constant (Hz) |
| m | multiplet |
| mg | milligram |
| M | molar |
| MHz | megahertz |
| mL | milliliter |
| mmol | millimol |
| NMR | Nuclear Magnetic Resonance |
| ORTEP | Oak Ridge Thermal Ellipsoid Plot Program |
| ppm | parts per million |
| % | percent |
| Ph | phenyl |
| q | quartet |
| s | singlet |
| sep | septet |
| t | triplet |
| TMS | tetramethylsilane |
| μL | microliter |

# Table of Contents

**1. Summary / Zusammenfassung**   1

   1.1 Summary   1

   1.2 Zusammenfassung   4

**2. Introduction**   7

**3. Overview of Thesis Results**   12

   3.1 Cycloaddition of a Cr-Cr quintuple bon   12

   3.2 Quintuple Bond Reactivity towards Group 16 and 17 Elements; Addition *vs*

   Insertion   12

   3.3 Reaction of a Cr-Cr Quintuple Bond with Phosphine Ligands   13

   3.4 Individual Contribution to Joint Publications   13

**4. Cycloaddition Reactions of a Chromium-Chromium Quintuple Bond**   16

   4.1 Introduction   16

   4.2 Results and Discussion   17

   4.3 Conclusion   23

   4.4 Experimental Section   24

   4.5 References   30

**5. Quintuple Bond Reactivity towards Group 16 and 17 Elements; Addition *vs***

   **Insertion**   34

   5.1 Introduction   35

   5.2 Results and Discussion   38

|  |  |
|---|---|
| 5.3 Conclusion | 47 |
| 5.4 Experimental Section | 50 |
| 5.5 References | 55 |
| **6. Reaction of a Cr-Cr Quintuple Bond with Phosphine Ligands** | **59** |
| 6.1 Introduction | 59 |
| 6.2 Results and Discussion | 60 |
| 6.3 Conclusion | 64 |
| 6.4 Experimental Section | 65 |
| 6.5 References | 68 |
| **7. List of Publications** | **71** |
| **8. Acknowledgments** | **72** |

# 1. Summary / Zusammenfassung

## 1.1 Summary

The objective of this work was to study the reactivity of a chromium-chromium quintuple bond complex towards a number of small molecules. The quintuply bonded chromium complex [Ap$^+$CrCrAp$^+$] (**1**) (where Ap$^+$H = 2,6-diisopropylphenyl-[6-(2,6-dimethylphenyl)-pyridin-2-yl]-amine) has been synthesized following a two-step reaction in which Ap$^+$H was deprotonated with n-BuLi and then reacted with CrCl$_2$. The obtained dimeric chloro-bridged complex [Ap$^+$CrCl(thf)]$_2$ was isolated and further reduced using KC$_8$ to give the quintuply bonded chromium complex **1**.

To extend an analogy between the Cr–Cr quintuple bond and the simple C–C double and triple bonds, reactions between quintuply bonded chromium dimers and substituted acetylenes and dienes have been explored. Thus, reacting **1** with alkynes and dienes led to a [2+2]-cycloaddition reaction in which the Cr–Cr bond order was reduced. This shows that the Cr–Cr quintuple bond can undergo similar cycloaddition reactions to those well known for C–C double and triple bonds.

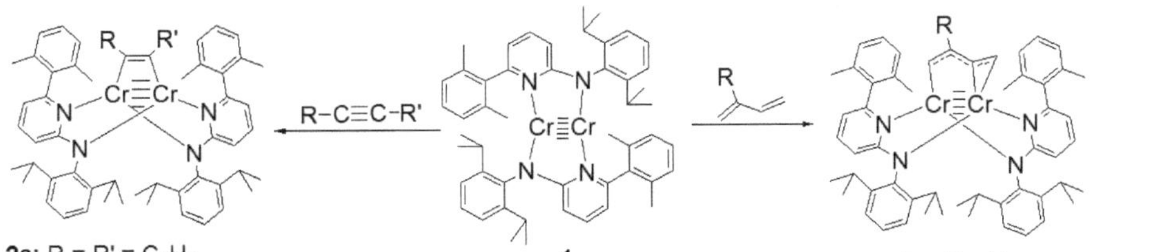

**2a:** R = R' = C$_6$H$_5$
**2b:** R = H, R' = C$_6$H$_5$
**2c:** R = H, R' = Si(CH$_3$)$_3$
**2d:** R = H, R' = C$_6$H$_4$-*p*-CH$_3$

**3a:** R = H
**3b:** R = CH$_3$

**Scheme 1:** Synthesis of the acetylene (**2**) and dienes (**3**) complexes.

Reactivity studies of the quintuple bond were extended to a variety of small inorganic molecules of group 16 and 17 elements. The dichromium platform can provide two to eight electrons to form complexes in which chromium can show an oxidation state from +II to +V. For group 16, $O_2$ gives selectively a dimeric $Cr^V$ species (**7**) while the other chalcogens led to addition reactions in which the $E_2^{2-}$ moiety binds to the $Cr^{II}$ unit (E = S, Se, Te) (**4**). For homodiatomic molecules of group 17, insertion of the quintuple bond into the corresponding X-X bond has been observed (X = Cl, Br, I) (**6**). Complex **1** was found to cleave the Se-Se and S-S bonds when reacted with diphenyldiselenide and diphenyldisulfide, respectively, to give the corresponding oxidative addition products (**5**).

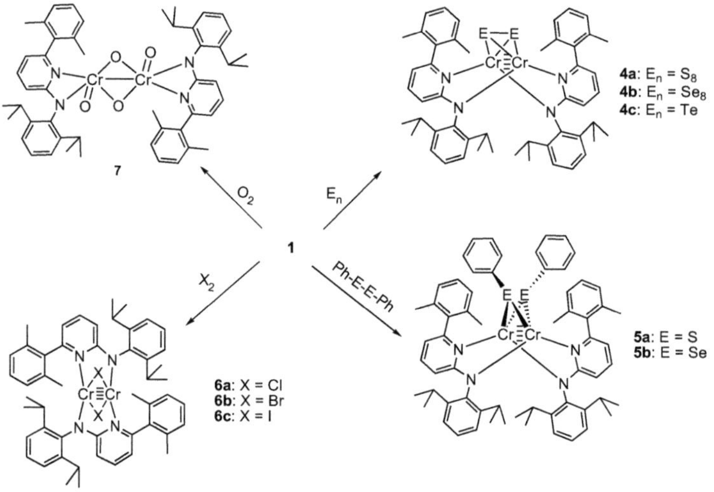

**Scheme 2:** Synthesis of complexes **4-7**.

The oxidative addition reactions of quintuple bond were further explored by reacting **1** with various phosphine ligands at room temperature to give the corresponding $Cr^{II}$ species. Selective oxidative addition resulting in the reduction of the bond order was observed.

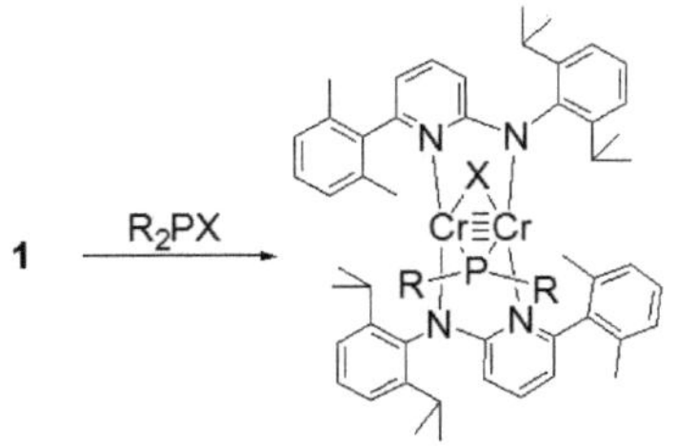

1 →[R₂PX]

**8a**: R = R = C₆H₅; X = H
**8b**: R = R = C₆H₅; X = Cl
**8c**: R = R = C₂H₅; X = Cl

**Scheme 3:** Synthesis of the phosphine complexes (**8**).

## 1.2 Zusammenfassung

Ziel dieser Arbeit waren Reaktivitätsuntersuchungen fünffachgebundener homobimetallischer Chrom-Komplexe gegenüber einer Anzahl kleiner Moleküle. Der fünffachgebundene Chrom-Komplex [Ap$^+$CrCrAp$^+$] (**1**) (mit Ap$^+$H = 2,6-Diisopropylphenyl-[6-(2,6-dimethylphenyl)-pyridin-2-yl]-amin) wurde in einer zweistufigen Reaktion synthetisiert, in welcher Ap$^+$H mit n-BuLi deprotoniert und mit CrCl$_2$ zur Reaktion gebracht wurde. Der dimere, Chlor-verbrückte Komplex [Ap$^+$CrCl(thf)]$_2$ wurde isoliert und mit KC$_8$ reduziert, um den fünffachgebundenen Chrom-Komplex **1** zu erhalten.

Um die Analogie zwischen Cr-Cr-Fünffachbindung und C-C-Doppel- und Dreifachbindung zu untersuchen, war ich an der Reaktion von fünffachgebundenen Chrom-dimeren mit substituierten Acetylenen und Dienen interessiert. So führte die Reaktion von **1** mit Alkinen und Dienen zu einer [2+2]-Cycloadditions-Reaktion, in welcher die Cr-Cr-Bindungsordnung reduziert wurde. Es zeigte sich, dass die Cr-Cr-Fünffachbindung vergleichbare Cycloadditionsreaktionen eingeht, wie dies für C-C-Doppel- und Dreifachbindung bereits gut bekannt ist.

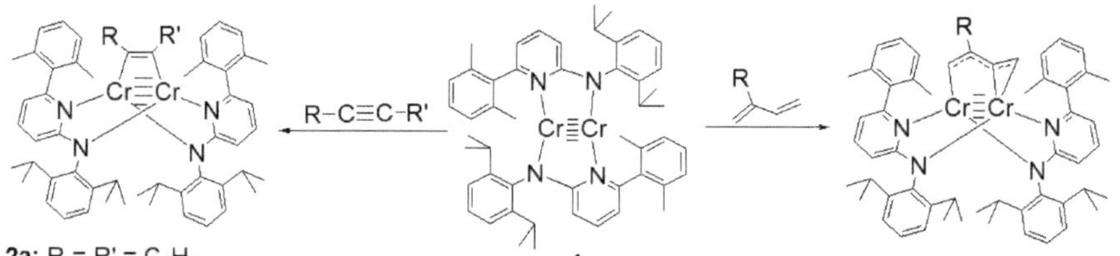

**2a:** R = R' = C$_6$H$_5$
**2b:** R = H, R' = C$_6$H$_5$
**2c:** R = H, R' = Si(CH$_3$)$_3$
**2d:** R = H, R' = C$_6$H$_4$-*p*-CH$_3$

**3a:** R = H
**3b:** R = CH$_3$

**Schema 1:** Synthese der Acetylen- (**2**) und Dien-Komplexe (**3**).

Die Reaktivitätsstudien an der Fünffachbindung wurden auf eine Vielzahl kleiner anorganischer Moleküle der Gruppe 16 und 17 ausgedehnt. Die Cr$_2$-Plattform liefert zwischen zwei und acht Elektronen, so dass Komplexe erzeugt werden können, in welchen Chrom in den Oxidationsstufen +II bis +V existieren kann. Von den Elementen der Gruppe 16 liefert $O_2$ selektiv eine dimere $Cr^V$-Spezies (**7**) während die anderen Chalcogene zu Additionsreaktionen führen, in denen das $E_2^{2-}$ Fragment an der $Cr^{II}$-Einheit bindet (E = S, Se, Te) (**4**).

Für homodiatomische Moleküle der Gruppe 17 wurde die Insertion der Fünffachbindung in die entsprechende X-X-Bindung beobachtet (**6**) (X = Cl, Br, I). Komplex **1** führt zur Bindungsspaltung von Se-Se- und S-S-Bindungen wenn er mit Diphenyldiselenid und Diphenyldisulfid umgesetzt wird, um die entsprechenden oxidativen Additionsprodukte (**5**) zu erhalten.

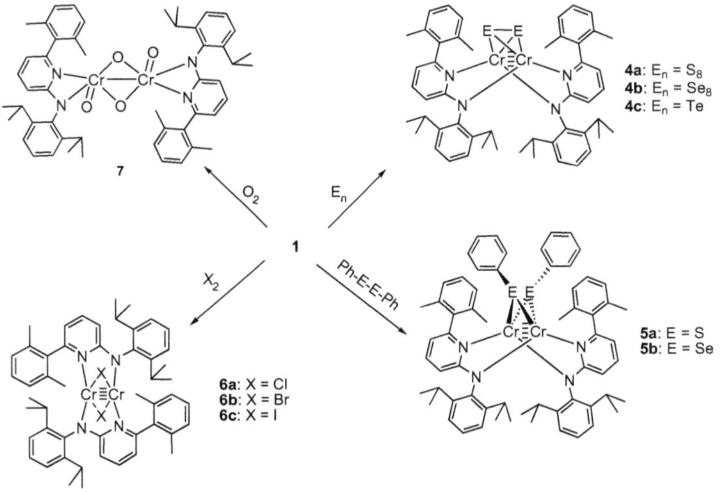

**Schema 2:** Synthese der Komplexe **4-7**.

Die oxidative Addition der Fünffachbindung wurde weiter erforscht, indem **1** mit verschiedenen Phosphin-Liganden bei Raumtemperatur umgesetzt wurde, um die entsprechenden $Cr^{II}$-Spezies zu erhalten. Es wurde eine selektive oxidative Addition im Zusammenhang mit einer Reduktion der Cr-Cr-Bindungsordnung beobachtet.

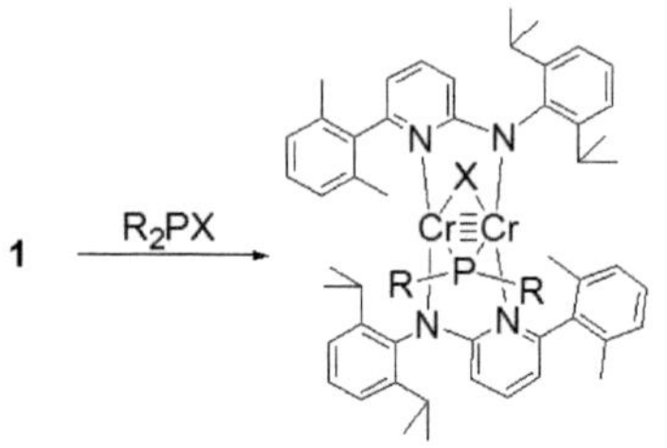

**Schema 3:** Synthese der Phosphin-Komplexe (**8**).

8a: R = R = C$_6$H$_5$; X = H
8b: R = R = C$_6$H$_5$; X = Cl
8c: R = R = C$_2$H$_5$; X = Cl

## 2. Introduction

The nature of chemical bond, the electronic structure and its reactivity is of fundamental interest.[1] In the case of a direct metal-metal bond between two transition elements, the most important interactions are between their d orbitals which can combine to form σ, π and δ orbitals. The first isolable compound featuring a δ bond was the quadruply bonded dirhenium compound $[Re_2Cl_8]^{2-}$ published in 1964.[2] Since then, many compounds with formally quadruply bonded transition metal atoms have been synthesized and explored in detail,[3] but the search for thermally stable and isolable dinuclear complexes with bond order higher than four has continued. In 2005, Power et al.[4] published the first Cr-Cr quintuple bond in the dimeric chromium compound Ar'CrCrAr' (where Ar' is the bulky $C_6H_3$-2,6-$(C_6H_3$-2,6-$iPr_2)_2)$, consisting of σ ($d_z^2$-$d_z^2$, $A_g$), 2 π ($d_{yz}$-$d_{yz}$, $d_{xz}$-$d_{xz}$, $A_u$, $B_u$) and 2 δ ($d_{x^2-y^2}$, $d_{xy}$-$d_{xy}$, $A_g$, $B_g$) bonds, with Cr-Cr distance of 1.8351(4) Å. Such accomplishment was possible by making the right choice of transition metal as well as by using a ligand assuring sufficient kinetic stabilization of the metal-metal bond. In this perspective, the group 6 elements are the best candidates for quintuple metal-metal bonds. Five out of their six valence electrons can be used to form the quintuple bond, leaving one electron free to share a bond with the surrounding ligand.[1-5] The choice of the ligand is the crucial parameter that keeps the transition metal in the lowest possible oxidation state, while maximizing the number of valence orbitals available for the formation of multiple bonds.[1-5] Since the metal-metal distance in Ar'CrCrAr' is slightly longer than the shortest Cr-Cr quadruple bond (1.828(2) Å) in the paddlewheel complex $[Cr_2(\mu-\eta^2-2-MeO-5-MeC_6H_3)_4]$,[6] the effects of differently substituted terphenyl ligands on the quintuple bond in arylchromium(I) dimers were also investigated. To this end, a series of complexes ArCrCrAr (Ar = $C_6H_2$-2,6-$(C_6H_3$-2,6-$iPr_2)_2$-4-R, where R = $SiMe_3$, OMe, and F) was synthesized, showing short Cr–Cr distances ranging from 1.8077(7) to 1.831(2) Å.[7] Continuous efforts were made to synthesize quintuply bonded complexes that have metal-metal bond lengths within the range of 1.68 Å (values for the Cr-Cr sextuple bond in $Cr_2$)[8] to 1.828 Å (shortest known Cr-Cr quadruple bond, see above). In 2007, Theopold et al. reported the dichromium complex [(μ-η$^2$-$^H$L$^{iPr}$)$_2$Cr$_2$], where $^H$L$^{iPr}$ = N,N'-bis(2,6-diisopropylphenyl)-1,4-diazabutadiene, with a Cr–Cr distance of 1.8028(9) Å.[9] Considering the potential of three-atom bridging ligands to form even shorter metal-metal distances (introduced as the Hein-Cotton principle),[3,10] two groups independently synthesized further examples of quintuply bonded dichromium complexes. In this regard, Tsai and co-workers [11] used amidinate ligands to form [Cr$_2$(μ-η$^2$-ArNC(R)NCAr)$_2$] compounds with Cr-Cr distances of 1.74 Å, while Kempe et al.[12] synthesized Cr$_2$ compound based on the aminopyridinate ligands with a Cr-Cr bond distance of 1.75 Å. The compounds mentioned above based on amidinates and aminopyridinates have almost the same [Cr$_2$(RNC(R')NR)$_2$] core unit but feature considerably different outer architectures (Figure 1).

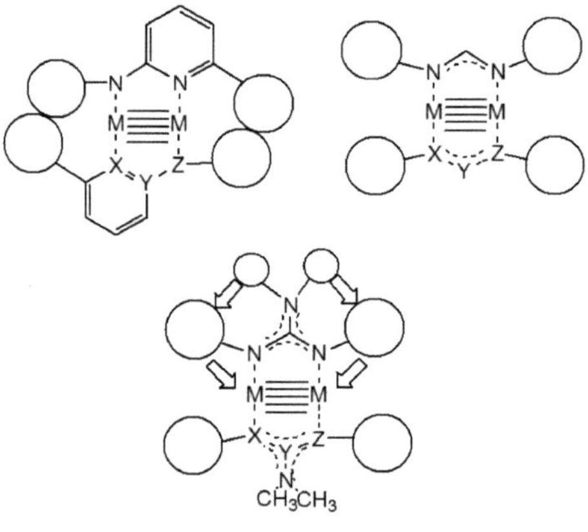

**Figure 1:** The role of the ligand in stabilising ultra-short metal-metal bonds: aminopyridinates (top left), amidinates (top right) and guanidinates (bottom).[13]

The two-wings-up arrangement observed for amidinates (Figure 1, top right) apparently causes much less inter-ligand repulsion within the bimetallic complex than the wing-up-wing-down arrangement observed for aminopyridinates (Figure 1, top left). This allows the generation of a closer N-C-N pincer and/or an alignment of nitrogen-atom-based orbitals that interact with chromium (lone pairs) towards each other, thus resulting in shorter Cr-Cr distances. Gradual closing of this pincer by applying steric pressure through the substituent at the bridging carbon atom should give rise to a further reduced distance between the two metal atoms. In this regard, the guanidinates seem to be suitable to achieve even shorter Cr-Cr bond lengths (Figure 1, bottom) by attaching different groups to the nitrogen not involved in the bonding with the metal. Thus, the shortest metal-metal bond observed to date (1.7293(12) Å) was obtained using guanidinate ligands.[14] Moreover, the isolation of quintuply bonded molybdenum complexes has been reported.[15] The new opportunity, now that stable compounds with such extreme bond orders are available, is to study quintuple bond reactivity and to understand these bonds chemically. For example, reactions of quintuply bonded systems with $N_2O$ and organic azides $RN_3$,[16] the carboalumination,[17] the activation of small molecules of group 15 elements,[18] the [2+2]- and [2+2+2]-cycloaddition reactions with alkynes[19] have been reported.

Within this work the quintuply bonded dichromium complex [Ap$^+$CrCrAp$^+$] (where Ap$^+$ = 2,6-diisopropylphenyl-[6-(2,6-dimethylphenyl)-pyridin-2-yl]-amine) has been synthesized[17] and then reacted with various small molecules. The isolated complexes have been characterized by X-ray diffraction studies, NMR spectroscopic investigations and elemental analysis.

**References**

[1] F. Wagner, A. Noor, R. Kempe, *Nat. Chem.* **2009**, *1*, 529-536.

[2] a) F. A. Cotton, N. F. Curtis, C. B. Harris, B. F. G. Johnson, S. J. Lippard, J. T. Mague, W. R. Robinson, J. S. Wood, *Science* **1964**, *145*, 1305-1307; b) F. A. Cotton, C. B. Harris, *Inorg. Chem.* **1965**, *4*, 330-333.

[3] F. A. Cotton, C. A. Murillo, R. A. Walton, in *Multiple Bonds Between Metal Atoms*, 3$^{rd}$ ed., Springer, New York **2005**.

[4] T. Nguyen, A. D. Sutton, M. Brynda, J. C. Fettinger, G. J. Long, P. P. Power, *Science* **2005**, *10*, 844-847.

[5] G. La Macchia, G. L. Manni, T. K. Todorova, M. Brynda, F. Aquilante, B. O. Roos, L. Gagliardi, *Inorg. Chem.* **2010**, *49*, 5216-5222.

[6] F. A. Cotton, S. A. Koch, M. Millar, *Inorg. Chem.* **1978**, *17*, 2084-2086.

[7] R. Wolf, C. Ni, T. Nguyen, M. Brynda, G. J. Long, A. D. Sutton, R. C. Fischer, J. C. Fettinger, M. Hellman, L. Pu, P. P. Power, *Inorg. Chem.* **2007**, *46*, 11277-11290.

[8] a) E. P. Kündig, M. Moskovits, G. A. Ozin, *Nature* **1975**, *254*, 503-504; b) W. Klotzbücher, G. A. Ozin, *Inorg. Chem.* **1977**, *16*, 984-987; c) Y. M. Efremov, A. N. Samoilova, L. V. Gurvich, *Opt. Spektrosc.* **1974**, *36*, 654-657; d) V. E. Bondybey, J. H. English, *Chem. Phys. Lett.* **1983**, *94*, 443-447.

[9] K. A. Kreisel, G. P. A. Yap, O. Dmitrenko, C. R. Landis, K. H. Theopold, *J. Am. Chem. Soc.* **2007**, *129*, 14162-14163.

[10] a) F. Hein, D. Tille, *Z. Anorg. Allg. Chem.* **1964**, *329*, 72-82; b) L. H. Gade, *Koordinationschemie*, Wiley-VCH, Weinheim **1998**.

[11] a) C.-W. Hsu, J.-S. K. Yu, C.-H. Yen, G.-H. Lee, Y. Wang, Y.-C. Tsai, *Angew. Chem.* **2008**, *120*, 10081-10084; *Angew. Chem. Int. Ed.* **2008**, *47*, 9933-9936; b) Y.-C. Tsai, C.-W. Hsu, J.-S. K. Yu, G.-H. Lee, Y. Wang, T.-S. Kuo, *Angew. Chem.* **2008**, *120*, 7250-7253; *Angew. Chem. Int. Ed.* **2008**, *47*, 7250-7253.

[12] A. Noor, F. R. Wagner, R. Kempe, *Angew. Chem.* **2008**, *120*, 7356-7359; *Angew. Chem. Int. Ed.* **2008**, *47*, 7246-7249.

[13] A. Noor, G. Glatz, R. Müller, M. Kaupp, S. Demeshko, R. Kempe, *Z. Anorg. Allg. Chem.* **2009**, *635,* 1149-1152.

[14] A. Noor, R. Kempe, *Chem. Rec.* **2010**, *10*, 413-416.

[15] a) Y.-C. Tsai, H.-Z. Chen, C.-C. Chang, J.-S. K. Yu, G.-H. Lee, Y. Wang, T.-S. Kuo, *J. Am. Chem. Soc.* **2009**, *131*, 12534-12535; b) S.-C. Liu, W.-L. Ke, J.-S. K. Yu, T.-S. Kuo, Y.-C. Tsai, *Angew. Chem.* **2012**, *124*, 6500-6503; *Angew. Chem. Int. Ed.* **2012**, *51*, 6394-6397.

[16] C. Ni, B. D. Ellis, G. J. Long, P. P. Power, *Chem. Commun.* **2009**, 2332–2334.

[17] A. Noor, G. Glatz, R. Müller, M. Kaupp, S. Demeshko, R. Kempe, *Nat. Chem.* **2009**, *1*, 322-325.

[18] C. Schwarzmaier, A. Noor, G. Glatz, M. Zabel, A. Y. Timoshkin, B. M. Cossairt, C. C. Cummins, R. Kempe, M. Sheer, *Angew. Chem.* **2011**, *123*, 7421-7424; *Angew. Chem. Int. Ed.* **2011**, *50*, 7283-7286.

[19] a) J. Shen, G. P. A. Yap, J.-P. Werner, K. H. Theopold, *Chem. Commun.* **2011**, *47*, 12191-12193; b) H.-Z Chen, S.-C. Liu, C.-H. Yen, J.-S. K. Yu, Y.-J. Shieh, T.-S. Kuo, Y.-C. Tsai, *Angew. Chem.* **2012**, *124*, 10488-10492; *Angew. Chem. Int. Ed.* **2012**, *51*, 10342-10346.

# 3. Overview of Thesis Results

This thesis comprises three publications, which are presented in chapters 4 to 6.

## 3.1 Cycloaddition Reactions of a Chromium-Chromium Quintuple Bond

In the first part, the reactivity of the quintuply bonded dichromium complex **1** with various alkynes and dienes has been discussed. An analogy has been drawn between quintuple bond and simple C–C double and triple bond as we observed cycloaddition reactions. A decrease in the chromium-chromium bond order has been observed indicating the oxidation of the $Cr_2$ moiety and reduction of the alkyne and diene ligands.

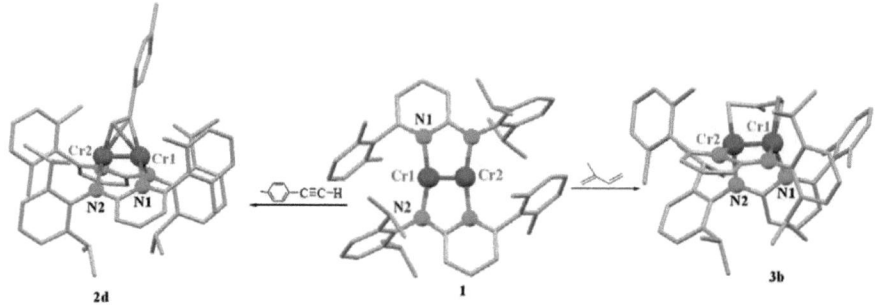

## 3.2 Quintuple Bond Reactivity towards Groups 16 and 17 Elements: Addition *vs* Insertion

The reactivity of quintuply bonded dichromium complex **1** was extended to various small molecules of group 16 and 17 elements. In the case of group 16, $O_2$ leads to the formation of dimeric $Cr^V$ specie. In contrast, higher homologues undergo addition reactions to give $Cr^{II}$ species. Thus, the quintuply bonded $Cr_2$ moiety preferentially donates two electrons. For homodiatomic molecules of group 17, insertion of the quintuple bond into the corresponding $X_2$ bond was observed (X = Cl, Br, I). The reaction of **1** with diphenyldisulfide or diphenyldiselenide shows the insertion of the Cr–Cr bond into the S–S or Se–Se bond and the formation of complexes in which the two Cr atoms are joined by two bridging PhS or PhSe groups.

# 3. Overview of Thesis Results

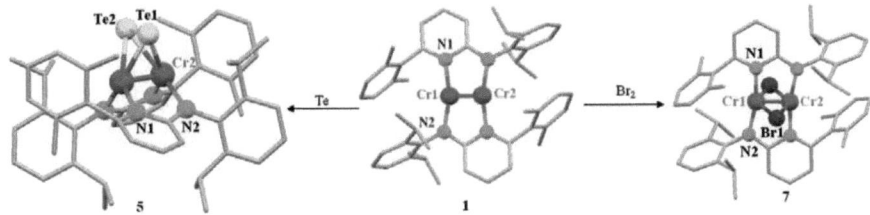

## 3.3 Reactions of a Cr-Cr Quintuple Bond with Phosphine Ligands

In this chapter the reaction behaviour of quintuply bonded chromium complex **1** with various phosphines was explored. The reaction of **1** with phosphines $R_2PX$ is selective and leads to the formation of phosphide-bridged chromium complexes, which have been characterized, by X-ray diffraction studies, NMR spectroscopic investigations and elemental analysis. The Cr-Cr bond length for the isolated compounds is typical for Cr-Cr quadruple bond.

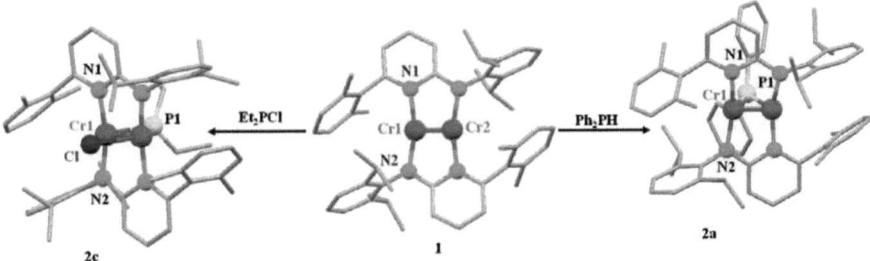

## 3.4 Individual contribution to joint publications

The results published in this thesis were obtained in collaboration with others and are published, submitted or are to be submitted as indicated below. In the following, contributions from all authors are specified. The asterisk indicates the corresponding author.

## 3.4.1 Chapter 4

This work was published in *Chem. Eur. J.* **2011**, *17*, 6900-6903, with the title '**Cycloaddition Reactions of a Chromium-Chromium Quintuple Bond.**'

Awal Noor, Emmanuel Sobgwi Tamne, Sadaf Qayyum, Tobias Bauer, Rhett Kempe*

Awal Noor and I studied the cycloaddition reactions, characterized all the data and wrote the manuscript.

I synthesized the quintuply bonded $Cr_2$ complex for the cycloaddition reactions.

Sadaf Qayyum helped to synthesize the $Ap^+H$ ligand.

Tobias Bauer did X-ray analyses of the compounds published in this work.

Rhett Kempe supervised this work and was involved in scientific discussions, suggestions and correction of the manuscript.

## 3.4.2 Chapter 5

This work has been accepted for publication in *Inorg. Chem.* with the title '**Quintuple Bond Reactivity towards Groups 16 and 17 Elements: Addition vs Insertion.**'

Emmanuel Sobgwi Tamne, Awal Noor, Sadaf Qayyum, Tobias Bauer, Rhett Kempe*

I synthesized and characterized the compounds of group 17 elements presented in this work.

Awal Noor and Sadaf Qayyum synthesized and characterized the compounds of group 16 elements and dichalcogenides presented in this work.

I wrote the manuscript with the help of Awal Noor and Rhett Kempe.

Tobias Bauer and Awal Noor did X-ray analyses of the compounds published in this work.

Rhett Kempe supervised this work and was involved in scientific discussions.

## 3.4.3 Chapter 6

This work is to be submitted to *Z. Anorg. Allg. Chem.* with the title '**Reactions of a Cr-Cr Quintuple Bond with Phosphine Ligands.**'

Emmanuel Sobgwi Tamne, Awal Noor, Tobias Bauer, Rhett Kempe*

I synthesized and characterized all the compounds presented in this work and the publication was written by me.

Awal Noor helped to carry out NMR experiments and was involved in the correction of the manuscript.

Tobias Bauer did X-ray analyses of the compounds published in this work.

Rhett Kempe supervised this work and was involved in scientific discussions, suggestions and correction of the manuscript.

# 4. Cycloaddition Reactions of a Chromium–Chromium Quintuple Bond

Awal Noor, Emmanuel Sobgwi Tamne, Sadaf Qayyum, Tobias Bauer, and Rhett Kempe*[a]

[a] Inorganic Chemistry II, University of Bayreuth, 95440 Bayreuth, Germany.
E-mail: kempe@uni-bayreuth.de

**Keywords**: alkynes, Chromium, cycloaddition, diene, multiple bonds, N-ligands.

**Published in:** *Chem. Eur. J.* 2011. *17*, 6900-6903.

**Abstract.** Addition of alkynes as well as dienes to a quintuply bonded dichromium complex was investigated and cycloaddition reactions were observed. The structures of the obtained compounds indicate the reduction of the alkyne and diene ligands, an oxidation of the chromium atoms, as well as the decrease of the Cr-Cr bond order.

## 4.1 Introduction

Bonds with unusually high bond orders have fascinated chemists for nearly half a century.[1] Molecules that have bond orders higher than four have been known for decades.[2] They can be found in transient diatomic molecules like $M_2$ (M = V, Nb, Cr, Mo).[3–6] The $Cr_2$ molecule, the most popular example among them, has been synthesized in the gas phase by means of pulsed photolysis[3] and by vaporization of the metal.[6] Furthermore, it can be isolated in an inert matrix.[4,5] Unfortunately, the stability restricts its use in synthesis (chemistry). In 2005, the Power group reported on the first stable molecule with a quintuply bonded dichromium unit;[7] for the definition of a quintuple bond please refer to a footnote in the paper. Shortly after, the Theopold group [8] reinitiated the "race" for the shortest metal–metal bond and many groups participated[9] in this still ongoing "hunt".[10] In 2009, the Tsai group reported on the first stable dimolybdenum complex with a quintuple bond.[11] The new opportunity, now that stable compounds with such extreme bond orders are in hand, is not to make shorter and shorter metal–metal bonds, but is to study quintuple bond reactivity and to understand these bonds chemically. Only very few reports on quintuple bond reactivity have appeared in the literature so far.[12] Quintuply bonded bimetallic complexes show a potential for small-molecule activation, particularly on a diatomic platform that can provide from two to eight (in principle even ten) electrons. These complexes feature not just low-valent metal centres, they are also coordinatively highly unsaturated. Alkyne and diene ligand have been successfully used as protecting ligands for low-valent Group 4 metallocenes by (for instance) the groups of Rosenthal, Erker, Beckhaus, and Mach.[13] If one uses a diatomic site of high bond order to activate such unsaturated molecules, cycloaddition reactions might be expected. We are interested in analogies of quintuple bonds to simple double and triple bonds. Here we report on reactions betweenchromium–chromium quintuple bond and substituted acetylenes and dienes.

## 4.2 Results and Discussion

The dichromium complex **1**[12a] reacts vigorously when acetylene is introduced into the reaction vessel and leads suddenly to black precipitation, which has been tentatively assigned to be polyacetylide material. However, **1** reacts smoothly in equimolar ratio with diphenylacetylene, phenylacetylene and trimethylsilylacetylene to give **2a**, **2b** and **2c**, respectively (Scheme 1). The formation of a 1:1 adduct is selective, even if **1** is treated with an excess of the corresponding acetylenes.

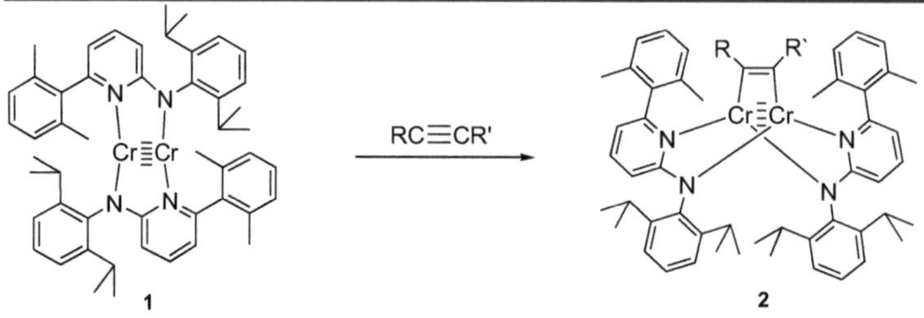

# 4. Cycloaddition Reactions of a Chromium-Chromium Quintuple Bond

2a: R = R` = C$_6$H$_5$
2b: R = H, R` = C$_6$H$_5$
2c: R = H, R` = Si(CH$_3$)$_3$
2d: R = H, R ' = C$_6$H$_4$-p-CH$_3$

**Scheme 1:** Synthesis of the acetylene complexes.

Compounds **2a–2c** are diamagnetic. The NMR spectra are in accordance with the nature (symmetry and the steric bulk) of the applied acetylenes. In the NMR spectra of complex **2a** we observe four doublets for the non-equivalent isopropyl CH$_3$ protons, two singlets for the methyl CH$_3$ protons and two septets for the isopropyl CH protons of the aminopyridinato ligands. NMR data remain (as expected) essentially unchanged when a solution of the complex in C$_7$D$_8$ was cooled to -50 °C. The $^{13}$C NMR spectrum shows the signal for carbon atoms of the coordinated diphenylacetylene at a chemical shift value of d = 197.9 ppm. The non-symmetric acetylene complexes **2b** and **2c** show (as expected) a different NMR pattern. Eight doublets for eight non-equivalent methyl protons of the isopropyl groups, four singlets for the methyl groups and four septets for the CH protons of the isopropyl groups were observed. The CH$_3$ protons of the SiMe$_3$ group were recorded as a sharp singlet at 0.57 ppm, pointing to the presence of one isomer only. NMR data of **2a–2c** are indicative that the aminopyridinato ligand is not flip-flopping (N-pyridine and N-amido atoms exchange positions) and the rotation of the 2,6-dialkylphenyl rings is frozen at room temperature. In consequence, a signal for each alkyl substituent of the 2,6-dialkylphenyl rings is observed for **2b** and **2c** and half of the signals for **2a** due to its symmetry. No reaction between bis(trimethylsilyl) acetylene and **1** is observed, even if the reaction mixture was heated (60 °C) for four hours. Since all these alkyne complexes resulted in weakly diffracting red/orange plates, we switched to p-tolylacetylene to synthesize **2d**. Crystals of **2d** partially suitable for X-ray analysis were grown by layering a solution of **2d** in toluene with hexane.[14] The NMR behavior of **2d** is quite similar to that observed for **2b**. The crystal structure of **2d** reveals a Cr-Cr distance of 1.8041(15) Å (Figure 1). It is a rather short quadruple bond and only just longer than the recently reported shortest quadruple bond.[15]

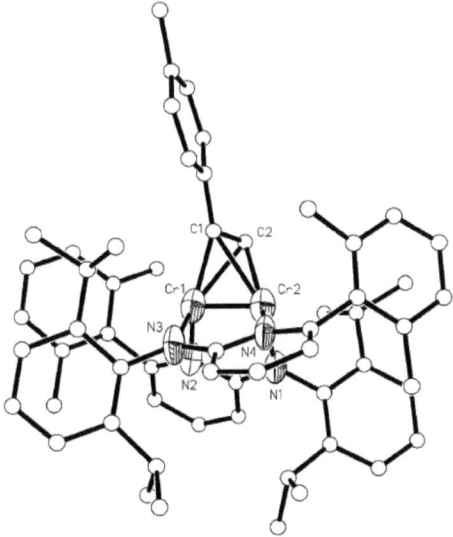

**Figure 1:** Molecular structure of **2d** - ORTEP representation (on the 50% probability level) for all non-carbon atoms. Hydrogen atoms have been deleted for clarity. Selected bond lengths [Å] and angles [°]: C1-C2 1.324(16), C1-Cr1 2.012(11), C1-Cr2 2.313(12),

C2-Cr2 2.011(10), C2-Cr1 2.297(11), N1-Cr2 2.020(6), N2-Cr1 2.008(6), N3-Cr1 2.021(7), N4-Cr2 2.025(6), Cr1-Cr2 1.8041(15); C2-C1-Cr1 84.3(8), C2-C1-Cr2 60.1(7), Cr1-C1-Cr2 48.7(3), C1-Cr1-C2 35.0(4), N2-Cr1-N3 104.9(3), N2-Cr1-C1 139.2(5), C1-Cr1-N3 115.6(5).

The p-tolylacetylene ligand coordinates perpendicularly to the two chromium atoms and the phenyl moiety is disordered, showing two possibilities of orientation (in 70:30 ratio). The C-C triple bond length of the coordinated acetylene (1.324(16) Å) is longer than the normal C-C triple bond length (1.181 Å) and close to the value of C=C bond (1.331 Å).[16] The elongation is indicative of a reduction of the bond order of the coordinated alkyne ligand.

Furthermore, we were interested to study the coordination of olefins. Compound **1** does not react with ethylene (1 bar), even if the reaction mixture was heated overnight at 80 °C, but it does react with dienes. One equivalent of buta-1,3-diene or isoprene was reacted with **1** to give the green complexes **3a** and **3b**, respectively, in good yields (Scheme 2). Crystals of **3a** and **3b** suitable for X-ray analysis were grown from hexane.[17] The coordinated dienes push the two aminopyridinato ligands downwards as the acetylenes do. Both dienes bind as cis isomers.

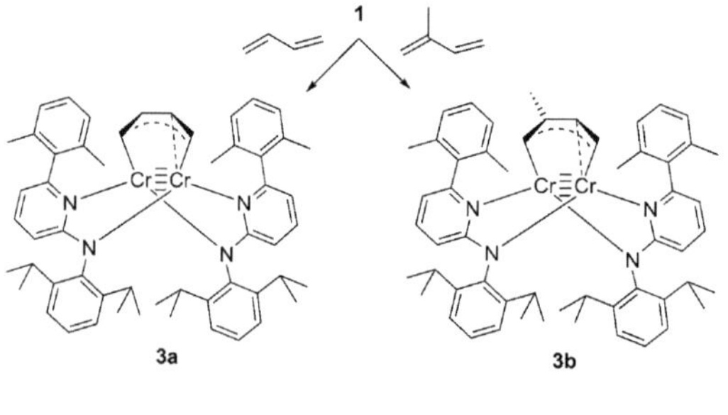

**Scheme 2:** Synthesis of the diene complexes.

The molecular structure of **3a** (Figure 2) is indicative of a μ,η,η²-coordination of the coordinated 1,3-butadiene to the two Cr atoms. The central C-C bond length of C1-C3 (1.427(8) Å) is longer than the two terminal C-C bonds, C1-C2 and C3-C4 with 1.405(8) and 1.388(8) Å, respectively. The coordinated isoprene in **3b** shows a μ,η,η²-coordination mode to the two Cr atoms (Figure 3) too. The Cr2-C4 distance (2.112(4) Å) is slightly shorter than that of Cr1-C1 (2.132(4) Å) and Cr1-C2 (2.165(4) Å). The C-C distances within the coordinated isoprene moiety (C1-C2 1.426(6), C2-C3 1.400(6) and C3-C4 (1.435(6) Å) are different from that of the coordinated butadiene. The central C-C bond length is shorter than the terminal C-C bond lengths, indicative of stronger multiple bond character in this part of the diene. In its extreme, resonating structure, a double bond between the central C atoms and a sigma-type bonding for the terminal C atoms can be assumed indicative of a Diels–Alder like addition of the dienes to the Cr-Cr quintuple bond. The Cr-Cr distances of **3a** (1.8227(14) Å) and **3b** (1.8228(8) Å) are equal. They are significantly longer than the quintuple bond in **1** (Cr1-Cr1A 1.750(1) Å)[12a] and they are in the expected range of a quadruple bond.[1] The slightly differences of the coordinated buta-1,3-diene versus isoprene might be caused by repulsion of the methyl groups of isoprene and the bulky aminopyridinato ligands. The $^1$H spectra of **3 a/b** are in accordance with their solid state structures and exclude the formation of any other isomers. Both **3a** and **3b** show hindered rotation of the 2,6-diaalkylphenyl rings in solution (at room temperature) and thus lead to eight doublets for $CH_3$ protons of the isopropyl groups, four singlets for the methyl protons and four septets for CH protons of the isopropyl group.

## 4. Cycloaddition Reactions of a Chromium-Chromium Quintuple Bond

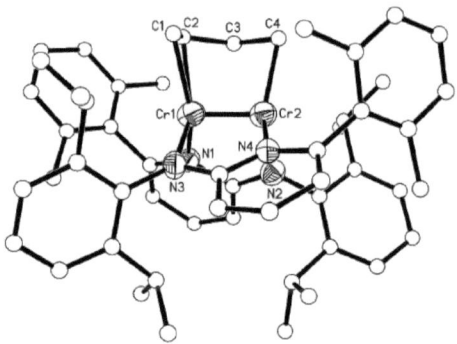

**Figure 2:** Molecular structure of **3a** ORTEP representation (on the 50% probability level) for all non-carbon atoms. Hydrogen atoms and one hexane molecule have been deleted for clarity. Selected bond lengths [Å] and angles [°]: C1-C2 1.405(8), C2-C3 1.427(8), C1-Cr1 2.179(6), C2-Cr1 2.119(6), C3-C4 1.388(8), C3-Cr2 2.432(6), C3-Cr1 2.469(7), C4-Cr2 2.091(6), N1-Cr1 2.052(5), N2-Cr2 2.038(5), N3-Cr1 2.057(5), N4-Cr2 1.992(5), Cr1-Cr2 1.8227(14); C1-C2-C3 120.4(6), C1-C2-Cr1 73.3(3), C3-C2-Cr1 85.9(4), C2-C1-Cr1 68.6(3), C4-C3-C2 122.5(6), C4- C3-Cr2 59.1(4), C2-C3-Cr2 97.6(4), C4-C3-Cr1 98.1(4), C2-C3-Cr1 58.9(3), Cr2-C3-Cr1 43.66(11), C3-C4-Cr2 86.2(4), Cr2-Cr1-N3 95.70(15), N1-Cr1-N3 103.73(18), Cr2-Cr1-C2 97.89(19), N1-Cr1-C2 109.1(2), N3-Cr1-C2 142.2(2), Cr2-Cr1-C1 99.2(2), N1-Cr1-C1 145.0(2), N3-Cr1-C1 104.8(2), C1-Cr1-C2 38.1(2), Cr2-Cr1-C3 67.09(18), N1-Cr1-C3 96.0(2),N3-Cr1-C3 155.6(2), C2-Cr1-C3 35.2(2), C1-Cr1-C3 63.5(2), Cr1-Cr2-N4 96.72(15), Cr1-Cr2-N2 95.00(16), N4-Cr2-N2 103.07(19), Cr1-Cr2-C4 99.73(19), N4-Cr2-C4 131.5(2), N2-Cr2-C4 120.4(2), Cr1-Cr2-C3 69.25(18), N4-Cr2-C3 148.7(2), N2-Cr2-C3 105.9(2), C4-Cr2-C3 34.7(2).

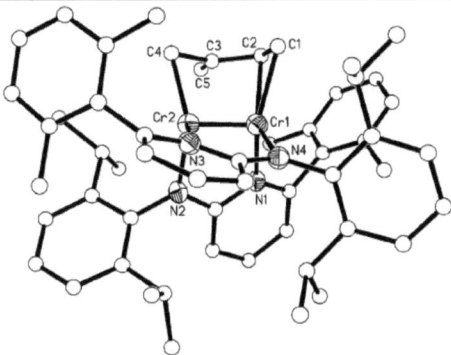

**Figure 3:** Molecular structure of **3b** ORTEP representation (on the 50% probability level) for all non-carbon atoms. Hydrogen atoms and one hexane molecule have been deleted for clarity. Selected bond lengths [Å] and angles [°]: C1-C2 1.426(6), C1-Cr1 2.132(4), C2-C3 1.40(6), C2-Cr1 2.165(4), C3-C4 1.435(6), C3-Cr2 2.539(4), C3-C5 1.513(4), C4-Cr2 2.112(4), N1-Cr1 2.074(3), N2-Cr2 2.050(3), N3-Cr2 1.984(3), N4-Cr1 2.055(3), Cr1-Cr2 1.8228(8); C2-C1-Cr1 71.9(2), N4-Cr1-N1 106.15(12), N4 Cr1-C1 102.00(15), N1-Cr1-C1 143.85(15), Cr2-Cr1-C2 98.09(12), N4-Cr1-C2 140.51(14), N1-Cr1-C2 108.87(15), C1-Cr1-C2 38.75(16), Cr1-Cr2-C4 99.37(12), N3-Cr2-C4 126.39(15), N2-Cr2-C4 127.43(16).

This hindered rotation is maintained in the $^{13}$C NMR spectrum as well. Complexes **3a/b** are not stable in solution at room temperature for longer periods of time, leading to oily material; however, they are very stable in solid form.

## 4.3 Conclusion

In summary, we present addition reactions of alkynes and dienes to a quintuply bonded dichromium complex. Both, alkynes and dienes undergo cycloaddition reactions in which the formal bond order of the Cr-Cr bond is reduced. Cr-Cr quintuple bonds can undergo similar (cycloaddition) reactions to those of well-understood C-C double and triple bonds. Considering the variety of reactions alkyne and diene complexes of Group 4 metallocenes can undergo,[13] we expect a rich chemistry of the cycloaddition product introduced here which we plan to study.

## 4.4 Experimental Section
### General Procedures

All manipulations were performed with rigorous exclusion of oxygen and moisture in Schlenk-type glassware on a dual manifold Schlenk line or in $N_2$ filled glove box (mBraun 120-G) with a highcapacity recirculator (< 0.1ppm O2). Solvents were dried by distillation from sodium wire / benzophenone. Complex **1** was prepared according to published procedure.[12a] Commercial $CrCl_2$ (Alfa Aesor) was used as received. Deuterated solvents were obtained from Cambridge Isotope Laboratories and were degassed, dried and distilled prior to use. NMR spectra were recorded on Varian 400 MHz at ambient temperature. The chemical shifts are reported in ppm relative to the internal TMS. Elemental analyses (CHN) were determined using a Vario EL III instrument. X-ray crystal structure analyses were performed by using a STOE-IPDS II equipped with an Oxford Cryostream low temperature unit. Structure solution and refinement was accomplished using SIR97,[18] SHELXL97[19] and WinGX.[20] CCDC-816309-11 contain the supplementary crystallographic data for this paper. These data can be obtained free of charge at www.ccdc.cam.ac.uk/conts/retrieving.html (or from the Cambridge Crystallographic Data Centre, 12 Union Road, Cambridge CB2 1EZ, UK; Fax: + 44-1223-336-033; e-mail: deposit@ccdc.cam.ac.uk).

# 4. Cycloaddition Reactions of a Chromium-Chromium Quintuple Bond

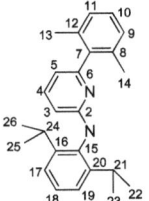

**Figure 4:** Labeling of the NMR signals.

**Synthesis of 2a:** Diphenylacetylene (0.018 g, 0.1 mmol) was added to **1** (0.082 g, 0.1 mmol) in toluene (10 mL) at room temperature. A sudden colour change to red was observed. The reaction mixture was stirred for three hours. The reaction mixture was allowed to stand overnight at room temperature to afford red crystalline material. Yield 0.082 g (82%). $C_{64}H_{68}Cr_2N_4$ (997.24): calcd. C 77.08, H 6.87, N 5.62; found C 77.07, H 6.93, N 5.58. $^1$H NMR (400 MHz, $C_6D_6$): $\delta$ = -0.65 (d, 6H, $J$ = 6.3 Hz, $H^{22/23/25/26}$), 1.08 (d, 6H, $J$ = 6.3 Hz, $H^{22/23/25/26}$), 1.13 (d, 6H, $J$ = 6.3 Hz, $H^{22/23/25/26}$), 1.38-1.41 (m, 12H, $H^{13,14,22/23/25/26}$), 2.24 (s, 12H, $H^{13,14}$), 3.50 (sep, 2H, $J$ = 6.3 Hz, $H^{21/24}$), 3.69 (sep, 2H, $J$ = 6.3 Hz, $H^{21/24}$), 5.74 (d, 2H, $J$ = 6.6 Hz, $H^3$), 6.23 (d, 2H, $J$ = 6.6 Hz, $H^5$), 6.26 (d, 2H, $J$ = 6.6 Hz, $H^{18}$), 6.48 (dd, 4H, $H^{Ph}$), 6.84 (dd, 4H, $H^{Ph,10}$), 7.07 (t, 2H, $J$ = 7.2 Hz, $H^4$), 7.18 (m, 4H, $H^{9,11/17,19}$), 7.27 (t, 4H, $J$ = 7.2 Hz, $H^{9,11/17,19}$), 7.58 (d, 4H, $J$ = 7.2 Hz, $H^{Ph}$) ppm. $^{13}$C NMR (100 MHz, $C_6D_6$): $\delta$ = 20.3 ($C^{13,14}$), 21.5 ($C^{13,14}$), 22.0 ($C^{22/23/25/26}$), 25.6 ($C^{22/23/25/26}$), 25.8 ($C^{22,23,/25,26}$), 26.4 ($C^{22,23/25,26}$), 27.8 ($C^{21,24}$), 28.1 ($C^{21,24}$), 108.0 ($C^3$), 109.1 ($C^5$), 124.3 ($C^{18}$), 125.5 (t, $C^{Ph}$), 127.1 ($C^{10}$), 127.5 ($C^{Ph}$), 128.5 ($C^{Ph}$), 134.0 ($C^{17,19}$), 135.5 ($C^{9,11}$), 136.1 ($C^{8,12}$), 136.6 (d,$C^{8,12}$), 137.2 ($C^4$), 137.6 ($C^7$), 144.1 ($C^{Ph}$), 144.2 ($C^{16,20}$), 145.0 ($C^{16,20}$), 145.3 ($C^{15}$), 156.1 ($C^6$), 169.8 ($C^2$), 197.9 ($C^{PhC=CPh}$) ppm.

**Synthesis of 2b:** Phenylacetylene (63 µL, 0.55 mmol) was added to **1** (0.246 g, 0.28 mmol) in toluene (5 mL) at room temperature. A sudden colour change to red was observed. The reaction mixture was stirred for half an hour, which resulted in the precipitation of the title compound. Solvent was filtered and allowed to further crystallize. Yield 0.227g (89%). $C_{58}H_{64}Cr_2N_4$ (921.15): calcd. C 75.63, H 7.00, N 6.08; found C 75.31, H 7.30, N 5.72. $^1$H NMR (400 MHz, $C_4D_8O$): $\delta$ = -0.38 (d, 3H, $J$ = 5.8 Hz, $H^{22/23/25/26}$), 0.23 (d, 3H, $J$ = 5.8 Hz, $H^{22/23/25/26}$), 0.86 (d, 3H, $J$ = 5.8 Hz, $H^{22723/25/26}$), 0.92 (d, 3H, $J$ = 5.8 Hz, $H^{22/23/25/26}$), 0.96 (d, 3H, $J$ = 5.8 Hz, $H^{22/23/25/26}$), 1.03 (s, 3H, $H^{13/14}$), 1.13 (d, 3H, $J$ = 5.8 Hz, $H^{22/23/25/26}$), 1.19 (s, 3H, $H^{13/14}$), 1.32 (two overlaping doublets, 6H, $H^{22/23/25,26}$), 1.88 (s, 3H, $H^{13/14}$), 1.97 (s, 3H, $H^{13/14}$), 3.23(br sep, 1H, $H^{21/24}$), 3.41 (br m, 3H, $H^{21,24}$), 5.90 (d, 1H, $J$ = 5.8 Hz, $H^3$), 5.97 (d, 1H, $J$ = 5.8 Hz, $H^3$), 6.16 (d, 1H, $J$ = 5.8 Hz, $H^5$), 6.22 (d, 1H, $J$ = 7.2 Hz, $H^5$), 6.39-6.64 (m, 6H, $H^{8/10/9/11/17/19}$), 6.86-6.91 (m, 2H, $H^{10/18}$), 7.07-7.38 (m, 9H, $H^{4/9/11/17/19/Ph}$), 7.55 (d, 2H, $J$ = 6.3 Hz, $H^{Ph}$), 11.33 (s, 1H, $H^{HC=CPh}$) ppm. $^{13}$C NMR (100 MHz, $C_6D_6$): $\delta$ = 19.8 ($C^{13,14}$), 21.1 ($C^{13,14}$), 22.4 ($C^{22/23/25/26}$), 23.6 ($C^{22/23/25/26}$), 25.0 ($C^{22,23,/25,26}$), 25.2 ($C^{22,23/25,26}$), 25.5 ($C^{22,23/25,26}$), 25.8 ($C^{22,23/25,26}$), 25.9 ($C^{22723/25/26}$), 27.8 ($C^{21,24}$), 27.9 ($C^{21,24}$), 28.5 ($C^{21/24}$), 29.4 ($C^{21/24}$), 107.2 ($C^3$), 107.6 ($C^3$), 108.7 ($C^5$), 108.8 ($C^5$), 123.9 ($C^{10}$), 124.0 ($C^{10}$), 125.4 ($C^{17,19}$), 125.6 ($C^{17,19}$), 126.5 ($C^{Ph}$), 127.3 ($C^{18}$), 127.4 ($C^{18}$), 127.9 ($C^{Ph}$), 128.5 ($C^{Ph}$), 134.0 ($C^{9/11}$), 134.3 ($C^{9/11}$), 136.0 ($C^{8,12}$), 136.1 ($C^{8,12}$), 136.2 (d,$C^{8,12}$), 135.8 ($C^4$), 137.2 ($C^7$), 137.3 ($C^7$), 144.4 ($C^{16/20}$), 144.5 ($C^{16/20}$), 144.7 ($C^{16/,20}$), 144.8 ($C^{16/,20}$), 144.9 ($C^{15}$), 145.0 ($C^{15}$), 155.7 ($C^6$), 155.8 ($C^6$), 169.0 ($C^2$), 169.6 ($C^2$), 183.3 ($C^{HC=CPh}$), 203 ($C^{HC=CPh}$) ppm.

**Synthesis of 2c:** Trimethylsilylacetylene (14 μL, 0.1 mmol) was added to **1** (0.082 g, 0.1 mmol) in toluene (10 mL) at room temperature. A sudden colour change to deep red was observed. The reaction mixture was stirred for half an hour at room temperature. Red crystalline material was obtained by cooling the solution, which was separated by filtration, and the filtrate was allowed to further crystallize. Yield 0.068 g (74%). $C_{55}H_{68}Cr_2N_4$ (916.40): calcd. C 72.02, H 7.47, N 6.11; found C 71.67, H 7.55, N 5.74. $^1$H NMR (400 MHz, $C_6D_6$): $\delta$ = 0.27 (d, 3H, $J$ = 6.8 Hz, $H^{22/23/25/26}$), 0.31 (d, 3H, $H^{22/23/25/26}$), 0.57 (s, 9H, $H^{SiMe3}$), 1.00 (d, 3H, $J$ = 6.8 Hz, $H^{22/23/25/26}$), 1.07 (d, 3H, $J$ = 6.8 Hz, $H^{22/23/25/26}$), 1.20 (s, 3H, $H^{13/14}$), 1.25 (s, 3H, $H^{13/14}$), 1.31 (d, 3H, $J$ = 6.8 Hz,

$H^{22/23/25/26}$), 1.33 (d, 3H, $J$ = 6.8 Hz, $H^{22/23/25/26}$), 1.41 (d, 6H, $J$ = 6.8 Hz, $H^{22/23/25,26}$), 1.43 (d, 3H, $J$ = 6.8 Hz, $H^{22/23/25/26}$), 1.90 (s, 3H, $H^{13/14}$), 2.10 (s, 3H, $H^{13/14}$), 3.49-3.60 (m, 2H, $H^{21/24}$), 3.71 (sep, 1H, $H^{21/24}$), 3.87 (sep, 1H, $J$ = 6.8 Hz, $H^{21/24}$), 5.65 (d, 1H, $J$ = 6.8 Hz, $H^3$), 5.72 (d, 1H, $J$ = 6.8 Hz, $H^3$), 6.21 (d, 1H, $J$ = 7.2 Hz, $H^5$), 6.26 (d, 1H, $J$ = 7.2 Hz, $H^5$), 6.40-6.61 (m, 6H, $H^{8/10/9/11/17/19}$), 6.77-6.84 (m, 2H, $H^{10/18}$), 6.98-7.01 (m, 4H, $H^4$), 7.17-7.22 (m, 4H, $H^{9/11/17/19}$), 12.51 (s, 1H, $H^{HC=SiMe3}$) ppm. $^{13}$C NMR (100 MHz, $C_6D_6$): $\delta$ = 1.7 ($C^{SiMe3}$), 19.1 ($C^{13,14}$), 19.4 ($C^{13,14}$), 19.6 ($C^{13,14}$), 21.3 ($C^{13,14}$), 23.6 ($C^{22/23/25/26}$), 23.7 ($C^{22/23/25/26}$), 23.8 ($C^{22/23/25/26}$), 25.1 ($C^{22/23,/25/26}$), 25.3 ($C^{22/23/25/26}$), 25.5 ($C^{22/23/25/26}$), 25.7 ($C^{22/23/25/26}$), 26.4 ($C^{22/23/25/26}$), 27.7 ($C^{21/24}$), 28.0 ($C^{21/24}$), 28.4 ($C^{21/24}$), 28.8 ($C^{21/24}$), 107.0 ($C^3$), 107.1 ($C^3$), 108.7 ($C^5$), 108.8 ($C^5$), 123.7 ($C^{10}$), 123.8 ($C^{10}$), 125.3 ($C^{17/19}$), 125.4 ($C^{17/19}$), 125.5 ($C^{17/19}$), 125.7 ($C^{17/19}$), 128.0 ($C^{18}$), 128.1 ($C^{18}$), 129.3 ($C^4$), 133.9 ($C^{9/11}$), 134.2 ($C^{9/11}$), 135.9 ($C^{8/12}$), 136.0 ($C^{8/12}$), 136.1 ($C^{8/12}$), 136.2 (d,$C^{8/12}$), 137.5 ($C^7$), 137.6 ($C^7$), 144.5 ($C^{16/20}$), 144.6 ($C^{16/20}$), 144.7 ($C^{16/,20}$), 144.8 ($C^{16/,20}$), 145.1 ($C^{15}$), 145.2 ($C^{15}$), 155.4 ($C^6$), 155.9 ($C^6$), 168.9 ($C^2$), 169.1 ($C^2$), 203.9 ($C^{HC=CSiMe3}$), 204 ($C^{HC=CSiMe3}$) ppm.

**Synthesis of 2d:** *p*-tolylacetylene (13 µL, 0.1 mmol) was added to **1** (0.082 g, 0.1 mmol) in toluene (10 mL) at room temperature. A sudden colour change to deep red was observed. The reaction mixture was stirred for half an hour at room temperature. Red crystalline material was obtained by layering with hexane and cooling the solution to -25°C. The material was separated by filtration and the filtrate was allowed to further crystallize. Yield 0.068 g (73%). $C_{59}H_{66}Cr_2N_4$ (935.17): calcd. C 75.78, H 7.11, N 5.99; found C 74.81, H 7.16, N 5.64. $^1$H NMR (400 MHz, $C_6D_6$): $\delta$ = -0.12 (d, 3H, $J$ = 6.8 Hz, $H^{22/23/25/26}$), 0.44 (d, 3H, , $J$ = 6.8 Hz $H^{22/23/25/26}$), 1.02 (d, 3H, $J$ = 6.8 Hz, $H^{22/23/25/26}$), 1.10 (d, 3H, $J$ = 6.8 Hz, $H^{22,23/25,26}$), 1.20 (s, 3H, $H^{13,14}$), 1.24 (d, 6H, $J$ = 6.8 Hz, $H^{22/23/25,26}$), 1.35 (s, 3H, $H^{13,14}$), 1.43 (d, 6H, $J$ = 6.8 Hz, $H^{22,23/25,26}$), 2.06 (s, 6H, $H^{13,14}$), 2.26 (s, 3H, $H^{HC=C6H4\text{-}CH3}$), 3.46 (sep, 1H, $J$ = 6.8 Hz, $H^{21/24}$), 3.53-3.61 (m, 2H, $H^{21/24}$), 3.76 (sep, 1H,

$J$ = 6.8 Hz, H$^{21/24}$), 5.68 (d, 1H, $J$ = 6.8 Hz, H$^3$), 5.74 (d, 1H, $J$ = 6.8 Hz, H$^3$), 6.20 (m, 1H, H$^{18}$), 6.26 (m, 1H, H$^{17/19}$), 6.26 (d, 1H, $J$ = 7.6 Hz, H$^5$), 6.30 (d, 1H, $J$ = 7.6 Hz, H$^5$), 6.42-6.49 (m, 4H, H$^5$), 6.55 (t, 1H, $J$ = 7.6 Hz, H$^4$), 6.78-6.80 (m, 2H, H$^{9/11}$), 6.95-7.01 (m, 5H, H$^{9/11,10,17,19}$), 7.26 (d, 2H, $J$ = 7.6 Hz, H C$^{6H4-CH3}$), 7.79 (d, 2H, $J$ = 7.6 Hz, H$^{C6H4-CH3}$), 11.63 (s, 1H, H$^{HC=C6H4-CH3}$) ppm. $^{13}$C NMR (100 MHz, C$_6$D$_6$): $\delta$ = 19.1 (C$^{CH3}$), 19.7 (d, C$^{13,14}$), 22.5 (d, C$^{22/23/25/26}$), 23.6 (d, C$^{22/23/25/26}$), 25.0 (C$^{22,23/25,26}$), 25.3 (C$^{22,23/25,26}$), 25.5 (C$^{22,23/25,26}$), 25.7 (C$^{22,23/25,26}$), 25.9 (C$^{22/23/25/26}$), 27.8 (C$^{21,24}$), 27.9 (C$^{21,24}$), 28.5 (C$^{21/24}$), 29.3 (C$^{21/24}$), 107.2 (C$^3$), 107.5 (C$^3$), 108.6 (C$^5$), 108.7 (C$^5$), 123.9 (C$^{10}$), 124.0 (C$^{10}$), 125.4 (C$^{17,19}$), 125.5 (C$^{17,19}$), 125.6 (C$^{Ph}$), 127.3 (C$^{18}$), 127.4 (C$^{18}$), 127.9 (C$^{Ph}$), 128.5 (C$^{Ph}$), 134.0 (C$^{9/11}$), 134.3 (C$^{9/11}$), 136.0 (C$^{8,12}$), 136.1 (C$^{8,12}$), 136.2 (d, C$^{8,12}$), 136.3 (C$^4$), 137.2 (C$^7$), 137.3 (C$^7$), 141.4 (C$^{p\text{-}Ph}$), 144.4 (C$^{16/20}$), 144.5 (C$^{16/20}$), 144.6 (C$^{16/20}$), 144.8 (C$^{16/20}$), 144.9 (C$^{15}$), 145.0 (C$^{15}$), 155.7 (C$^6$), 155.8 (C$^6$), 169.0 (C$^2$), 169.5 (C$^2$), 182.1 (C$^{HC=CPh}$), 202 (C$^{HC=CPh}$) ppm.

**Synthesis of 3a:** 1,3 butadiene (excess) was introduced into **1** (0.2 g, 0.244 mmol) in toluene (20 mL) at -40°C as colour started to change to brown green. The solution was stirred for 20 minutes at this temperature and then allowed to warm to room temperature and further stirred for half an hour. The clear brown green solution was cooled to afford brown crystals of the title compound. Crystals suitable for X-ray solution were grown from hexane solution. Yield 0.152g (65 %). C$_{54}$H$_{64}$Cr$_2$N$_4$·C$_6$H$_{12}$ (957.26): calcd. C 75.28, H 8.00, N 5.85; found C 75.28, H 8.03, N 6.19. $^1$H NMR (400 MHz, C$_6$D$_6$, 298 K): $\delta$ = -0.25 (dd, 1H, $J$ = 2.8 Hz, $J$ = 8.8 Hz, H$^{Butadiene}$), 0.73 (s, 3H, H$^{13/14}$), 0.76 (d, 3H, $J$ = 6.6 Hz, H$^{22/23/25/26}$), 0.82 (br d, 1H, H$^{Butadiene}$), 0.90 (d, 3H, $J$ = 6.6 Hz, H$^{22/23/25/26}$), 0.93 (d, 3H, $J$ = 6.6 Hz, H$^{22/23/25/26}$), 1.07 (d, 3H, $J$ = 6.6 Hz, H$^{22/23/25/26}$), 1.22 (d, 3H, $J$ = 6.6 Hz, H$^{22/23/25/26}$), 1.26 (d, 3H, $J$ = 6.6 Hz, H$^{22/23/25/26}$), 1.43 (m, 6H, H$^{13/14/22,23/25,26}$), 1.46 (d, 3H, $J$ = 6.6 Hz, H$^{22/23/25/26}$), 1.64 (s, 3H, H$^{13/14}$), 1.91 (s, 3H, H$^{13/14}$), 2.39 (sep, 1H, $J$ = 6.6 Hz, H$^{21/24}$), 3.55-3.65 (m, 2H, H$^{21/24,butadiene}$), 3.71 (sep, 1H, $J$ = 6.6 Hz, H$^{21/24}$), 4.43 (sep, 1H,

$J$ = 6.6 Hz, H$^{21,24}$), 5.53 (m, 1H, H$^{Butadiene}$), 5.63-5.68 (m, 2H, H$^3$ ), 6.15 (d, 1H, H$^{9/11/17/19}$), 6.28 (dd, 2H, $J$ = 5.8 Hz, $J$ = 9.1 Hz, H$^5$), 6.41 (br dd, 2H, H$^{9/11}$ ), 6.51 (d, 1H, $J$ = 7.2 Hz, H$^{9/11}$ ), 6.60-6.66 (m, 2H, H$^4$), 6.60 (dd, 1H, $J$ = 7.5 Hz, H$^{10/18}$), 6.78 (d, 1H, $J$ = 7.2 Hz, H$^{9/11}$ ), 6.81 (br, 1H, H$^{17/19}$), 7.05-7.32 (m, 5H, H$^{10/18,17/19}$) ppm. $^{13}$C NMR (C$_6$D$_6$, 298 K): $\delta$ = 18.1 (C$^{13/14}$), 18.8 (C$^{13/14}$), 19.5 (C$^{13/14}$), 20.0 (C$^{13/14}$), 22.4 (C$^{22/23/25/26}$), 23.7 (C$^{22/23/25/26}$), 24.3 (C$^{22/23/25/26}$), 24.7 (C$^{22/23/25/26}$), 25.7 (C$^{22/23/25/26}$), 26.1 (C$^{22/23/25/26}$), 26.4 (C$^{22/23/25/26}$), 26.8 (C$^{22/23/25/26}$), 27.4 (C$^{21/24}$), 28.0 (C$^{21/24}$), 28.2 (C$^{21/24}$), 28.4 (C$^{21/24}$), 45.1 (C$^{Butadiene}$), 47.8 (C$^{Butadiene}$), 48.0 (C$^{Butadiene}$), 99.2 (C$^{Butadiene}$), 107.5 (C$^3$), 108.4 (C$^3$), 108.8 (C$^5$), 109.0 (C$^{17/19/9/11}$), 109.8 (C$^5$), 124.0 (C$^{10/18}$), 124.4 (C$^{10/18}$), 125.3 (C$^{10/18}$), 125.7 (C$^{17/19}$), 125.8 (C$^{17/19}$), 126.1 (C$^{17/19}$), 126.3 (C$^{17/19}$), 126.9 (C$^{9/11}$), 127.0 (C$^{9/11}$), 127.3 (C$^{9/11}$), 127.4 (C$^{9/11}$), 133.2 (C$^4$), 133.7 (C$^4$), 134.9 (C$^{8/12}$), 135.7 (C$^{8/12}$), 138.3 (C$^{8/12}$), 138.5 (C$^{8/12}$), 138.8 (C$^7$), 139.7 (C$^7$), 144.5 (C$^{16/20}$), 145.3 (C$^{16/20}$), 145.7 (C$^{16/20}$), 145.7 (C$^{16/20}$), 145.8 (C$^{15}$), 146.5 (C$^{15}$), 156.0 (C$^6$), 157.0 (C$^6$), 170.6 (C$^2$), 172.6 (C$^2$) ppm.

**Synthesis of 3b:** Isoprene (10 µL, 0.1 mmol) was added to **1** (0.082 g, 0.1 mmol) in toluene (5 mL) at room temperature as colour started to change to brown green. The solution was shaken for 10 minutes and then allowed to stay at low temperature for 24 h. Toluene was evaporated and hexane (10 mL) was added resulting in sudden precipitation of the title compound. The product was separated by filtrations and filtrate was allowed to further afford the product. Yield 0.072g (74%). C$_{61}$H$_{80}$Cr$_2$N$_4$ (973.31): calcd. C 75.27, H 8.28, N 5.76; found C 74.68, H 7.91, N 5.86. $^1$H NMR (400 MHz, C$_6$D$_6$, 298 K): $\delta$ = -1.06 (d, 1H, $J$ = 3.3 Hz), 0.38 (s, 3H, H$^{CH3}$), 0.73 (d, 3H, $J$ = 6.6 Hz, H$^{22/23/25/26}$), 0.86-092 (m, 7H, H$^{22,23/25,26,isoprene}$), 1.17-1.27 (m, 9H, H$^{13/14/22,23/25,26}$), 1.38-1.51 (m, 9H, H$^{13/14/22,23/25,26}$), 1.87 (s, 3H, H$^{13/14}$), 2.15-2.22 (m, 4H, H$^{13/14,isoprene}$), 2.38 (sep, 1H, $J$ = 6.8 Hz, H$^{21/24}$), 3.37 (dd, 1H, $J$ = 6.8 Hz, H$^{Isoprene}$), 3.51 (sep, 1H, $J$ = 6.8 Hz, H$^{21/24}$), 3.62 (sep, 1H, $J$ = 6.8 Hz, H$^{21/24}$), 4.66 (t, 1H$^{Isoprene}$, $J$ = 9.10 Hz), 4.78 (sep,

1H, $J = 6.8$ Hz, $H^{21,24}$), 5.62 (d, 1H, $J = 6.8$ Hz, $H^3$), 5.76 (d, 1H, $J = 6.8$ Hz, $H^5$), 6.14 (d, 1H, $J = 8.8$ Hz, $H^3$), 6.24 (d, 1H, $J = 8.8$ Hz, $H^5$), 6.28 (d, 1H, $J = 6.2$ Hz, $H^{9/11}$), 6.37 (d, 1H, $J = 6.2$ Hz, $H^{9/11}$), 6.48 (dd, 1H, $J = 6.2$ Hz, $J = 8.8$ Hz, $H^4$), 6.53 (d, 1H, $J = 6.2$ Hz, $H^{9/11}$), 6.60 (dd, 1H, $J = 6.2$ Hz, $J = 8.8$ Hz, $H^4$), 6.66 (dd, 1H, $J = 7.5$ Hz, $H^{10/18}$), 6.83 (m, 2H, $H^{9/11,17/19}$), 7.10-7.32 (m, 4H, $H^{10/18,17/19}$) ppm. $^{13}$C NMR ($C_6D_6$, 298 K): $\delta =$ 14.3 ($C^{22/23/25/26}$), 18.6 ($C^{CH3}$), 18.6 ($C^{13/14}$), 19.3 ($C^{13/14}$), 20.5 ($C^{Isoprene}$), 22.0 ($C^{13/14}$), 22.4 ($C^{22/23/25/26}$), 23.0 ($C^{22/23/25/26}$), 23.2 ($C^{22/23/25/26}$), 24.6 ($C^{22/23/25/26}$), 25.0 ($C^{13/14}$), 25.1 ($C^{22/23/25/26}$), 25.6 ($C^{22/23/25/26}$), 26.5 ($C^{22/23/25/26}$), 27.4 ($C^{21/24}$), 27.9 ($C^{21/24}$), 28.0 ($C^{21/24}$), 28.5 ($C^{21/24}$), 31.9 ($C^{Isoprene}$), 38.4 ($C^{Isoprene}$), 52.4 ($C^{Isoprene}$), 89.2 ($C^{Isoprene}$), 107.2 ($C^3$), 109.0 ($C^3$), 110.5 ($C^5$), 110.8 ($C^5$), 124.2 ($C^{10/18}$), 124.6 ($C^{10/18}$), 124.8 ($C^{17/19}$), 126.0 (d, $C^{17/19}$), 126.4 ($C^{17/19}$), 127.0 ($C^{17/19}$), 127.4 ($C^{9/11}$), 127.9 ($C^{9/11}$), 128.1 ($C^{9/11}$), 128.4 ($C^{9/11}$), 132.2 ($C^4$), 133.6 ($C^4$), 134.5 ($C^{8/12}$), 136.2 ($C^{8/12}$), 137.6 ($C^{8/12}$), 138.1 ($C^{8/12}$), 138.2 ($C^7$), 139.7 ($C^7$), 144.9 ($C^{16/20}$), 145.8 ($C^{16/20}$), 146.0 ($C^{16/,20}$), 146.4 ($C^{16/20}$), 146.5 ($C^{15}$), 147.1 ($C^{15}$), 156.9 ($C^6$), 156.8 ($C^6$), 170.7 ($C^2$), 173.1 ($C^2$) ppm.

## Acknowledgements

Financial support from the Deutsche Forschungsgemeinschaft (DFG KE 756/20-1) is gratefully acknowledged. E.S.T. thanks the Deutscher Akademischer Austausch Dienst (DAAD) for a Ph.D. scholarship. We also thank Dr. Germund Glatz for his support in the X-ray lab.

## 4.5 References

[1] a) F. Wagner, A. Noor, R. Kempe, *Nat. Chem.* **2009**, *1*, 529-536; b) F. A. Cotton, L. A. Murillo, R. A. Walton, *Multiple Bonds Between Metal Atoms*, 3rd ed., Springer, Berlin **2005**.

[2] M. D. Morse, *Chem. Rev.* **1986**, *86*, 1049-1109.

[3] Yu. M. Efremov, A. N. Samoilova, L. V. Gurvich, *Opt. Spektrosk.* **1974**, *36*, 654-657.

[4] E. P. Kündig, M. Moskovits, G. A. Ozin, *Nature* **1975**, *254*, 503-504.

[5] W. Klotzbücher, G. A. Ozin, *Inorg. Chem.* **1977**, *16*, 984-987.

[6] V. E. Bondybey, J. H. English, *Chem. Phys. Lett.* **1983**, *94*, 443-447.

[7] T. Nguyen, A. D. Sutton, M. Brynda, J. C. Fettinger, G. J. Long, P. P. Power, *Science* **2005**, *310*, 844-847.

[8] K. A. Kreisel, G. P. A. Yap, O. Dmitrenko, C. R. Landis, K. H. Theopold, *J. Am. Chem. Soc.* **2007**, *129*, 14162-14163.

[9] a) R. Wolf, C. Ni, T. Nguyen, M. Brynda, G. J. Long, A. D. Sutton, R. C. Fischer, J. C. Fettinger, M. Hellman, L. Pu, P. P. Power, *Inorg. Chem.* **2007**, *46*, 11277-11290; b) Y.-C. Tsai, C.-W. Hsu, J.-S. K. Yu, G.-H. Lee, Y. Wang, T.-S. Kuo, *Angew. Chem.* **2008**, *120*, 7360-7363; *Angew. Chem. Int. Ed.* **2008**, *47*, 7250-7253; c) A. Noor, F. R. Wagner, R. Kempe, *Angew. Chem.* **2008**, *120*, 7356-7359; *Angew. Chem. Int. Ed.* **2008**, *47*, 7246-7249; d) C.-W. Hsu, J.-S. K. Yu, C.-H. Yen, G.-H. Lee, Y. Wang, Y.-C. Tsai, *Angew. Chem.* **2008**, *120*, 10081-10084; *Angew. Chem. Int. Ed.* **2008**, *47*, 9933-9936; e) A. Noor, G. Glatz, R. Müler, M. Kaupp, S. Demeshko, R. Kempe, *Z. Anorg. Allg. Chem.* **2009**, *635*, 1149-1152.

[10] A. Noor, R. Kempe, *Chem. Rec.* **2010**, *10*, 413-416.

[11] Y.-C. Tsai, H.-Z. Chen, C.-C. Chang, J.-S. K. Yu, G.-H. Lee, Y. Wang, T.-S. Kuo, *J. Am. Chem. Soc.* **2009**, *131*, 12534-12535.

[12] a) A. Noor, G. Glatz, R. Müler, M. Kaupp, S. Demeshko, R. Kempe, *Nat. Chem.* **2009**, *1*, 322-325; b) C. Ni, B. D. Ellis, G. J. Long, P. P. Power, *Chem. Commun.* **2009**, 2332-2334.

[13] For selected reviews, please see a) U. Rosenthal, V. V. Burlakov in *Titanium and Zirconium in Organic Synthesis* (Ed.: I. Marek), Wiley-VCH, Weinhiemk, **2002**, pp. 355-389; b) U. Rosenthal, V. V. Bulakov, M. A. Bach, T. Beweries, *Chem. Soc. Rev.* **2007**, *36*, 719-728; c) U. Rosenthal, V. V. Burlakov, P. Arndt, W. Baumann, A. Spannenberg, V. B. Shur, *Eur. J. Inorg. Chem.* **2004**, 4739-4749; d) U. Rosenthal, V. V.

## 4. Cycloaddition Reactions of a Chromium-Chromium Quintuple Bond

Burlakov, P. Arndt, W. Baumann, A. Spannenberg, *Organometallics* **2003**, *22*, 884-900; e) U. Rosenthal, P. -M. Pellny, F. G. Kirchbauer, G. Frank, V. V. Burlakov, *Acc. Chem. Res.* **2000**, *33*, 119 –129; f) G. Erker, G. Kehr, R. Fröhlich, *Coord. Chem. Rev.* **2006**, *250*, 36– 46; g) G. Erker, G. Kehr, R. Fröhlich, *J. Organomet. Chem.* **2004**, *689*, 4305-4318; h) G. Erker, G. Kehr, R. Fröhlich, *Adv. Organomet. Chem.* **2004**, *51*, 109-162; i) G. Erker, *Chem. Commun.* **2003**, 1469-1476; j) G. Erker, *Acc. Chem. Res.* **2001**, *34*,309-317; k) G. Erker, C. Krüger, G. Müller, *Adv. Organomet. Chem.* **1985**, *24*, 1-39; For selected key papers please see: l) M. Lamač, A. Spannenberg, H. Jiao, S. Hansen, W. Baumann, P. Arndt, U. Rosenthal, *Angew. Chem.* **2010**, *122*, 2999-3002; *Angew. Chem. Int. Ed.* **2010**, *49*, 2937-2940; m) P. M. Pellny, F. G. Kirchbauer, V. V. Burlakov, W. Baumann, A. Spannenberg, U. Rosenthal, *J. Am. Chem. Soc.* **1999**, *121*, 8313-8323; n) A. Ohff, P. Kosse, W. Baumann, A. Tillack, R. Kempe, H. Goerls, V. V. Burlakov, U. Rosenthal, *J. Am. Chem. Soc.* **1995**, *117*, 10399-10400; o) J. Karl, G. Erker, R. Fröhlich, *J. Am. Chem. Soc.* **1997**, *119*, 11165-11173; p) B. Temme, G. Erker, J. Karl, H. Luftmann, R. Fröhlich, S. Kotila, *Angew. Chem.* **1995**, *107*, 1867-1869; *Angew. Chem. Int. Ed. Engl.* **1995**, *34*, 1755-1757; q) G. Erker, J. Wicher, K. Engel, F. Rosenfeldt, W. Dietrich, C. Krüger, *J. Am. Chem. Soc.* **1980**, *102*, 6344-6346; r) S. Kraft, E. Hanuschek, R. Beckhaus, D. Haase, W. Saak, *Chem. Eur. J.* **2005**, *11*, 969-978; s) I. M. Piglosiewicz, R. Beckhaus, W. Saak, D. Haase, *J. Am. Chem. Soc.* **2005**, *127*, 14190-14191; t) S. Kraft, R. Beckhaus, D. Haase, W. Saak, *Angew. Chem.* **2004**, *116*, 1609-1614; *Angew. Chem. Int. Ed.* **2004**, *43*, 1583-1587; u) M. Horáček, P. Stepnicka, J. Kubista, R. Gyepes, K. Mach, *Organometallics* **2004**, *23*, 3388-3397; v) V. Kupfer, U. Thewalt, I. Tislerova, P. Stepnicka, R. Gyepes, J. Kubista, M. Horáčcek, K. Mach, *J. Organomet. Chem.* **2001**, *620*, 39-50; w) J. Hiller, U. Thewalt, M. Polášek, L. Petrusová, V. Varga, P. Sedmera, K. Mach, *Organometallics* **1996**, *15*, 3752-3759.

[14] P$\bar{1}$, a = 10.2540(7), b = 12.5920(8), c = 22.5750(17) Å; α = 78.123(6), β = 88.772(6), γ = 74.336(5)° and R$^1$ = 0.1020 [I>2σ (I)]; wR$^2$ = 0.3006 (all data).

[15] S. Horvath, S. I. Gorelsky, S. Gambarotta, I. Korobkov, *Angew. Chem.* **2008**, *120*, 10085-10088; *Angew. Chem. Int. Ed.* **2008**, *47*, 9937-9940.

[16] F. H. Allen, O. Kennard, D. G. Watson, L. Brammer, A. G. Orpen, R. Taylor, *J. Chem. Soc. Perkin Trans. 2* **1987**, S1.

[17] a) $P\bar{1}$, a = 11.0270(7), b = 12.3620(7), c = 22.2350(13) Å; α = 106.759(5), β = 95.199(5), γ = 93.004(5)° and $R_1$=0.0509 [I>2σ(I)]; $wR^2$ = 0.1098 (all data); b) $P2_1/n$, a = 12.6960(6), b = 10.9090(5), c = 38.5910(19) Å; β = 92.628(4)° and $R_1$ = 0.0661 [I>2σ(I)]; $wR^2$ = 0.1615 (all data).

[18] A. Altomare, M. C. Burla, M. Camalli, G. L. Cascarano, C. Giacovazzo, A. Guagliardi, A. G. G. Moliterni, G. Polidori, R. Spagna, SIR 97: A new tool for crystal structure determination and refinement. *J. Appl. Cryst.* **1999**, *32*, 115-119.

[19] SHELX97 Programs for crystal structure analysis (Release 97-2). (G. M. Sheldrick, Institut für Anorganische Chemie der Universität, Tammanstrasse 4, D-3400 Göttingen, Germany, **1998**).

[20] L. J. Farrugia, WinGX suite for small-molecule single-crystal crystallography. *J. Appl. Cryst.* **1999**, *32*, 837-838.

## 5. Quintuple Bond Reactivity towards Group 16 and 17 Elements; Addition vs Insertion

Emmanuel Sobgwi Tamne, Awal Noor, Sadaf Qayyum, Tobias Bauer, and Rhett Kempe*[a]

[a] Inorganic Chemistry II, University of Bayreuth, 95440 Bayreuth, Germany.
E-mail: kempe@uni-bayreuth.de

**Keywords**: Addition reactions, chromium, insertion reactions, multiple bonds, N-ligands.
**Accepted for publication in:** *Inorg. Chem.*

**Abstract.** The low valent, coordinatively unsaturated, and formally quintuply bonded bimetallic aminopyridinato chromium complex **1** was investigated regarding its reactivity towards group 16 and 17 elements. Reaction of **1** with $O_2$ yielded a dimeric Cr oxo complex **2**, a compound with a high formal oxidation state carrying both bridging and terminal oxo ligands. Reactions with the higher homologues of the group lead to the formation of dimeric $Cr^{II}$ complexes in which $E_2^{2-}$ ligands were formed [E = S (**3**), Se (**4**) and Te (**5**)]. Here the quintuply bonded dichromium unit formally undergoes an addition reaction. Reaction of **1** with the homo diatomic molecules of the group 17 elements leads to products in which the Cr-Cr quintuple bond is inserted into the corresponding $X_2$ molecule [X = Cl (**6**), Br (**7**) and I (**8**)]. Complex **1** was also found to insert into the S–S and Se–Se bonds of 1,2-diphenyldisulfane or the corresponding selenium compound (complexes **9** and **10**, respectively). All the compounds have been characterized by NMR and elemental analysis. Additionally, eight of the complexes have been characterized by X-ray analysis. The bimetallic $Cr^{II}$ complexes feature metal-metal distances between 1.8369(18) and 1.918(12) Å.

# 5. Quintuple Bond Reactivity towards Group 16 and 17 Elements; Addition vs. Insertion

## 5.1 Introduction

Bond orders are of fundamental interest in chemistry.[1] Looking at the simple hydrocarbons ethane, ethylene and acetylene we see a drastic increase in reactivity with an increasing bond order. Formally, we store electrons in the C-C linkage and can use them for additional bond formations. Of course, it is not that simple. For the related dinitrogen compounds the triply bonded molecule, rich in electrons too, is stable like a rock and the other two, diazene and hydrazine, are the more reactive. Very high bond orders, namely quintuple and sextuple bonds can be observed between transition metals.[2] Molecules having such high bond orders are known for decades,[3] the transient diatomic molecules $M_2$ (M = V, Nb, Cr, Mo) being prominent examples.[4] Unfortunately, their instability and highly demanding and partially highly unselective synthesis restricts their use in (for instance) inorganic synthesis, small molecule activation or catalysis. In 2005, the group of Power reported a breakthrough in this field, the synthesis of the first stable molecule having a quintuple bond.[5] Shortly after and inspired by that compound the groups of Theopold, Tsai and us reported on N-ligand stabilized dichromium complexes having a quintuple bond.[6] Ultrashort metal-metal bond distances have been observed for these compounds.[1,7] The record is right now at 1.73 Å.[8] Meanwhile, a considerable number of quintuply bonded dichromium complexes have been reported.[9] Furthermore, related dimolybdenum compounds were synthesized.[10] With these stable compounds in hand we are enabled to study their reactivity and by doing so we may understand quintuple bonds chemically. We have previously shown that quintuply bonded complexes can provide from two to eight electrons and observed the carboalumination of a quintuple bond as well as its oxidation with $O_2$, during which $Cr^I$ has been oxidized to $Cr^V$ (Scheme 1).[11a]

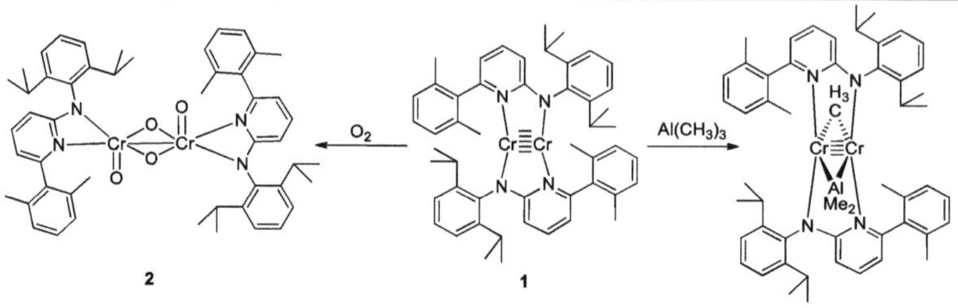

# 5. Quintuple Bond Reactivity towards Group 16 and 17 Elements; Addition vs. Insertion

**Scheme 1:** Carboalumination and oxidation of a chromium-chromium quintuple bond.

At the same time, Power et al. have shown that the ligand role is crucial in quintuple bond reactivity by observing complete cleavage of the quintuple bond for complexes of the type ArCrCrAr (Ar = substituted terphenyl) if reacted with $N_2O$ and azides (Scheme 2).[12]

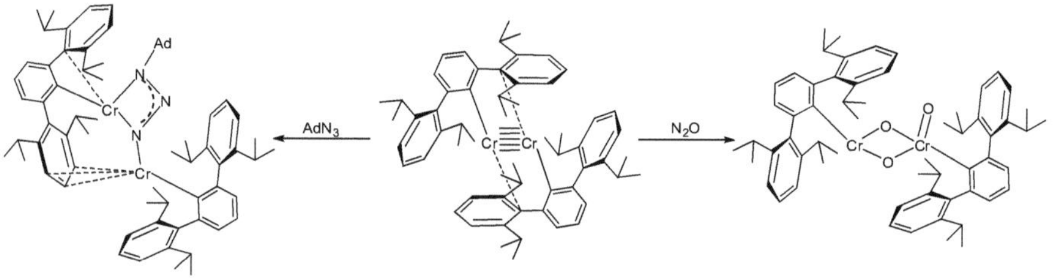

**Scheme 2:** Cleavage of a quintuple bond by $N_2O$ and admantyl azide (AdN$_3$).

Cycloaddition reactions were observed by Theopold et al. and us, when alkynes and dienes were reacted with quintuply bonded dichromium complexes (Scheme 3).[11b,13] While extending our understanding of quintuple bond reactivity we have shown that molecules like phosphorus, yellow arsenic, and AsP$_3$ can be activated under mild conditions and in a highly selective manner.[11c]

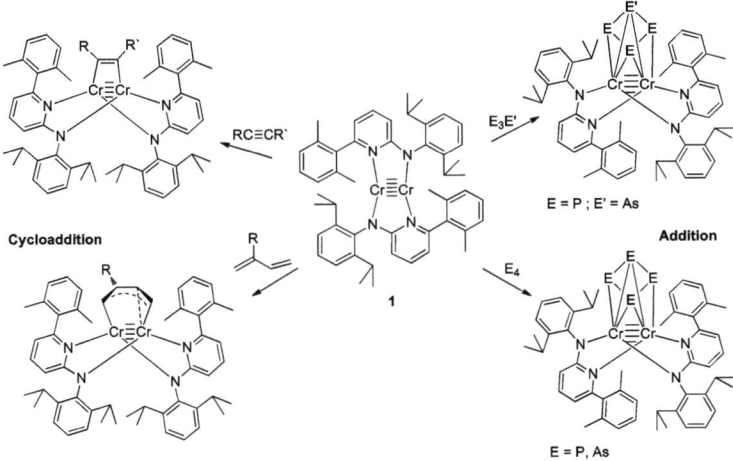

**Scheme 3:** Cycloaddition and addition reactions of quintuply bonded Cr complexes.

Tsai et al. have shown that amidinate ligands stabilized complexes with a Cr-Cr bond length of 1.7404(8) Å[6d] show weak coordination of THF and 2-MeTHF (2-MeTHF = 2-methyltetrahydrofuran) with long Cr-O distances of 2.579(4) and 2.305(7) Å, respectively (Scheme 4).[9] The ligated THF and 2-MeTHF ligands are labile and result in comprehensive elongation of the Cr-Cr bond to 1.8115(12) and 1.7636(5) Å, respectively.

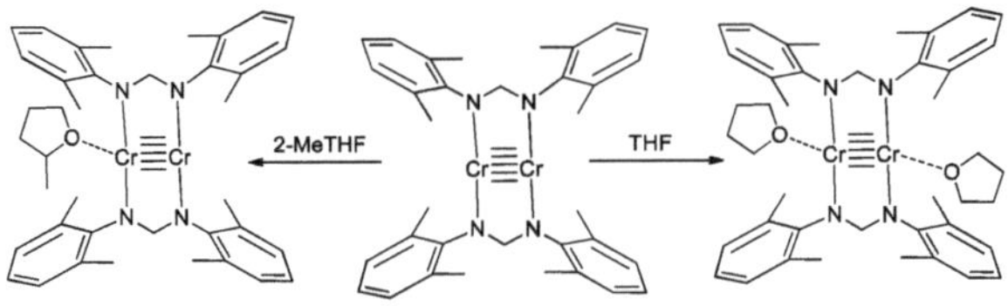

**Scheme 4:** Weak coordination of THF and 2-MeTHF to quintuply bonded Cr complexes.

## 5. Quintuple Bond Reactivity towards Group 16 and 17 Elements; Addition vs. Insertion

Complexes stabilized by diamidopyridine ligands undergo two electrons oxidation if reacted with AgOTf and thus completely cleave the Cr-Cr bond.[9] Similar cleavage of the quintuple bond has been observed when the diamidopyridine ligand stabilized Cr complexes were reacted with [18] crown-6 ether (Scheme 5).[9]

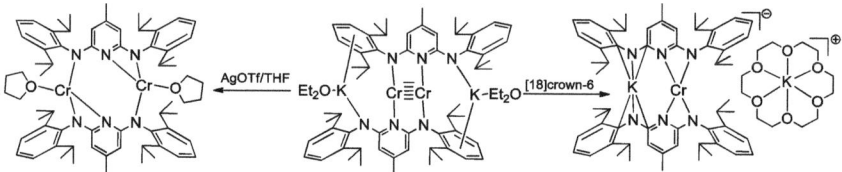

**Scheme 5:** Quintuple bond reactivity of diamidopyridine stabilized Cr complexes.

Very recently, Tsai et al. have extended the reactivity studies of quintuply bonded complexes towards Mo and have observed [2+2+2] cycloaddition reactions with alkynes.[14]

Here we report our results involving reactions between chromium–chromium quintuple bond and chalcogens (S, Se and Te), halogens and cleavage of single bonds between chalcogenides (PhSSPh and PhSeSePh).

### 5.2 Results and Discussion

Complex **1** was readily prepared following the published procedure.[11a] Reacting **1** with excess of oxygen afforded **2** as red brown solid in reasonable isolated yield of 72% (Scheme 6).[11a]

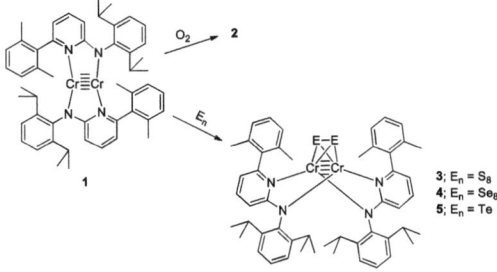

**Scheme 6:** Synthesis of **2-5**.

$^1$H NMR shows **2** to be a diamagnetic compound with sharp signals between 0 and 8 ppm. We observe two doublets for the isopropyl groups, two singlets for the CH$_3$ protons and two septets for the CH proton of the isopropyl-groups. Crystals suitable for X-ray analysis were grown from hexane solution. The solid state structure shows **2** to be a dimeric Cr(V) oxo complex, in which two oxo ligands are bridging and two oxo ligands coordinate terminally (Figure 1). Chromium oxo complexes of this type are rare [15] and were first reported by Herberhold and co-workers.[16] The bridging Cr–O bond distances of 1.791(2) Å is similar to the respective 1.817(4) Å distance in the Cr(V) complex [CpCr(O)(µ-O)]$_2$.[16] The terminal Cr–O bond length of 1.563(2) Å is considerably shorter than the bridging Cr–O distances and indicates a strong double bond character. It is also shorter than the Cr$^V$ complex of Herberhold (1.594(3) Å).[16] The long Cr–Cr distance of 2.5314(10) Å corresponds to a single bond between the two metal centres.

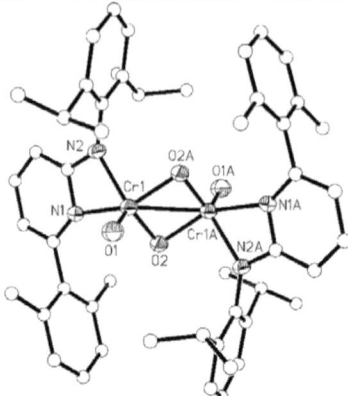

**Figure 1:** Molecular structure of **2**. ORTEP representation on the 50% probability level for all non-carbon atoms. Hydrogen atoms have been deleted for clarity. Selected bond lengths [Å] and angles [°]: Cr1–N1 2.053(2), Cr1–N2 1.946(2), O1–Cr1 1.563(2), O2–Cr1 1.791(2), Cr1–Cr1A 2.5314(10) Å; Cr1–Cr1A–O1A 121.69(9), Cr1–Cr1A–O2 45.02(7), N2–Cr1–N1 65.20(10)°.

Impressed by the highly selective oxidation of the quintuple bond by $O_2$ we became interested in exploring the higher homologues of that group. Complex **1** reacts with $S_8$ at room temperature and with $Se_8$ and Te at 60 °C in toluene to afford **3**, **4** and **5**, respectively, in high yields (Scheme 6). Complexes **3-5** are diamagnetic and have been characterized by solution NMR spectroscopy and elemental analysis. The $^1$H NMR spectra of **3-5** in $C_6D_6$ are as expected for diamagnetic complexes, featuring slight shifting for a multitude of sharp resonances between 0 and 8 ppm. We observe four doublets for the non-equivalent isopropyl $CH_3$ protons, two singlets for the methyl $CH_3$ protons and two septets for the isopropyl CH protons of the aminopyridinato ligands. Additionally, **3** and **5** have also been characterized by X-ray analysis. The solid state structure shows the oxidation of $Cr_2$ moiety to give structurally similar bimetallic $\mu,\eta^2$-disulfide (**3**) and $\mu,\eta^2$-ditelluride (**5**) complexes (Figure 2). Quintuply bonded diatomic molecules can provide from two to ten electrons if activated with small molecules and can thus show a large disparity in the formal oxidation states of chromium. It can be seen that for dimeric oxo complex **2**, $Cr^I$ has been oxidized to $Cr^V$ and for **3-5** to $Cr^{II}$. In **3** the S-S distance of 2.058(4) Å is shorter than expected for disulfur ligands,[17] but it lies close to other known bimetallic chromium complexes [2.028(2) Å].[18] The averaged Cr-S bond distance, 2.3885(3) Å is longer than those known for relevant $\mu,\eta^2$-disulfide bimetallic chromium complexes.[17,18,19,20] In comparison to $\mu,\eta^2$-disulfide ligands, $\mu,\eta^2$-ditelluride ligands on diatomic transition metals platform are rare and unknown for chromium, thus **5** represents the first example of such Cr complexes. The Te-Te bond distance of 2.6878(8) Å is comparable to a closely known ditellurido vanadium (IV) complex [2.6961(5) Å][21] and ditellurido iron complexes [2.700-2.719(4) Å][22], and are normal for a ditelluride single bond. The Cr-Te distances lie in the range of 2.7231(14) to 2.7511(13) Å. The Cr-Cr bond distances of 1.847(2) Å in **3** and 1.8369(18) Å in **5** lie in the range known for quadruply bonded complexes (Table 1).

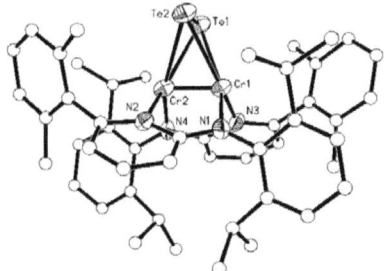

**Figure 2:** Molecular structure of **5**, ORTEP representation on the 50% probability level for all non-carbon atoms. Hydrogen atoms have been deleted for clarity.

## 5. Quintuple Bond Reactivity towards Group 16 and 17 Elements; Addition vs. Insertion

**Table 1:** Selected bond lengths [Å] and angles [°] for complexes **3** and **5**.

| 3 (E = S) | | 5 (E = Te) | |
|---|---|---|---|
| Cr-Cr | 1.847(2) | Cr-Cr | 1.8369(18) |
| $N_{Am}$-Cr | 2.004(8) | $N_{Am}$-Cr | 2.030(7) |
| $N_{Py}$-Cr | 2.004(8) | $N_{Py}$-Cr | 2.030(7) |
| E-E | 2.058(4) | E-E | 2.6878(18) |
| E-Cr | 2.388(3) | E-Cr | 2.7351(13) |
| E-E-Cr | 64.48(11) | E-E-Cr | 60.57(3) |
| Cr-E-Cr | 45.48(7) | Cr-E-Cr | 39.24(4) |
| Cr-Cr-$N_{Am}$ | 95.5(2) | Cr-Cr-$N_{Am}$ | 95.13(19) |
| Cr-Cr-$N_{Py}$ | 97.4(3) | Cr-Cr-$N_{Py}$ | 97.79(19) |
| $N_{Am}$-Cr-$N_{Py}$ | 104.7(3) | $N_{Am}$-Cr-$N_{Py}$ | 106.2(3) |
| Cr-Cr-E | 67.66(9) | Cr-Cr-E | 70.36(6) |
| E-Cr-E | 51.03(10) | E-Cr-E | 58.85(3) |

$N_{Am} = N_{Amido}$, $N_{Py} = N_{Pyridine}$

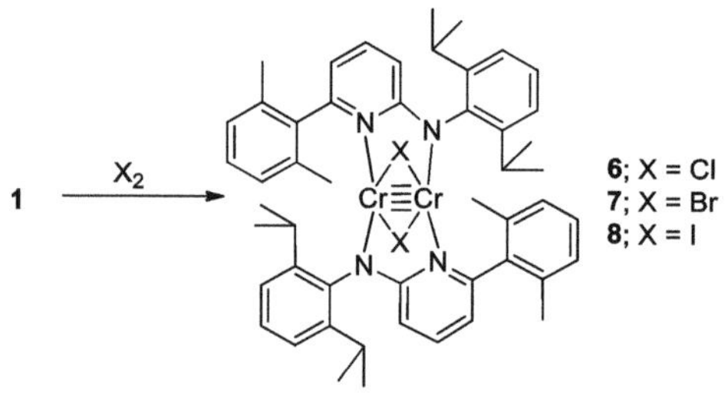

**Scheme 7:** Synthesis of **6-8**.

Furthermore, we were interested to study the reactivity of **1** towards homo diatomic group 17 molecules, especially in comparison to the activation of sulfur, selenium and tellurium. Stirring toluene/benzene solution of **1** with the corresponding $X_2$ molecule yielded complexes **6-8** in good yields (**6**: X = Cl; **7**: X = Br; **8**: X = I) (Scheme 7). Compounds **6-8** are diamagnetic and their $^1$H NMR spectra in solution show a characteristic peaks pattern of two doublets for the non-equivalent $CH_3$ protons of the isopropyl groups, one singlet for the $CH_3$ protons and one septet for the CH proton of the isopropyl groups. A single set of signals was observed for the aromatic protons. Solid-state structure analyses show halogen bridged bimetallic $Cr^{II}$ complexes. We observed selective oxidative addition even if an excess of $Cl_2$, $Br_2$ or $I_2$ was used. In similar reactions Power et al. have observed not only complete cleavage of the quintuple bond but also excess of $I_2$ led to form $[CrI_2(thf)_2]$ species.[23] Structurally characterized chromium complexes with bridging iodides are also rare and only a few examples are known.[24] The halogen-bridged complexes **6-8** are structurally similar and feature Cr atoms coordinated to five other atoms thus forming a distorted trigonal bipyramidal geometry (Figure 3). The Cr-Cr distances of 1.918(12) Å for **6**, 1.868(18) Å for **7** and 1.874(4) Å for **8** are typical for quadruple bonds. The average Cr-X bond length could be classified as Cr-Cl (2.438 Å) < Cr-Br (2.575 Å) < Cr-I (2.758 Å). The short Cr-Cl bond could be due to the more electronegative nature of chlorine than bromine and iodine. This has also an impact on the values of the Cr-X-Cr angles which amount to 46.32(3), 42.54(4) and 39.74(7)° for **6**, **7** and **8**, respectively. It is also in accordance with the values of the Cr-Cr bond distance, which is longest in **6** [1.918(12) Å] (Table 2). The longer $Cr-N_{py}$ distances compared to $Cr-N_{am}$ bond distances show the localization of the anionic function on the amido N-atom, which is typical of an amidopyridine mode of coordination of the ligands.[25] Details of the X-ray crystal structure analyses are summarized in Table 2.

# 5. Quintuple Bond Reactivity towards Group 16 and 17 Elements; Addition vs. Insertion

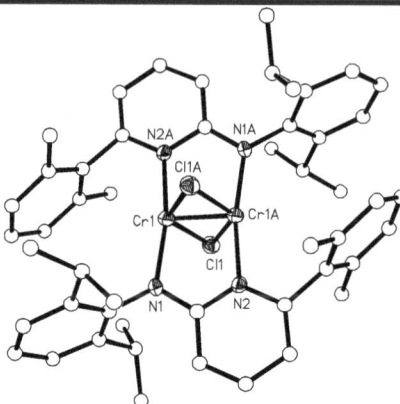

**Figure 3:** Molecular structure of **6**. ORTEP representation on the 50% probability level for all non-carbon atoms. Hydrogen atoms have been deleted for clarity.

**Table 2:** Selected bond lengths [Å] and angles [°] for complexes **6**, **7** and **8**.

| 6 (X = Cl) | | 7 (X = Br) | | 8 (X = I) | |
|---|---|---|---|---|---|
| Cr-Cr | 1.918(12) | Cr-Cr | 1.868(18) | Cr-Cr | 1.874(4) |
| $N_{Am}$-Cr | 1.918(12) | $N_{Am}$-Cr | 1.997(5) | $N_{Am}$-Cr | 2.004(8) |
| $N_{Py}$-Cr | 2.023(2) | $N_{Py}$-Cr | 2.003(5) | $N_{Py}$-Cr | 2.028(9) |
| Cr-X | 2.438(13) | Cr-X | 2.574(12) | Cr-X | 2.757(2) |
| Cr-Cr-X | 67.61(3) | Cr-Cr-X | 69.73(6) | Cr-Cr-X | 69.81(11) |
| Cr-X-Cr | 46.32(5) | Cr-X-Cr | 42.54(4) | Cr-X-Cr | 39.71(3) |
| Cr-Cr-$N_{Am}$ | 97.91(7) | Cr-Cr-$N_{Am}$ | 98.64(14) | Cr-Cr-$N_{Am}$ | 97.5(3) |
| Cr-Cr-$N_{Py}$ | 92.55(8) | Cr-Cr-$N_{Py}$ | 93.25(14) | Cr-Cr-$N_{Py}$ | 94.2(3) |
| $N_{Am}$-Cr-$N_{Py}$ | 169.54(19) | $N_{Am}$-Cr-$N_{Py}$ | 168.11(19) | $N_{Am}$-Cr-$N_{Py}$ | 168.3(4) |
| $N_{Am}$-Cr-X | 91.76(7) | $N_{Am}$-Cr-X | 91.22(14) | $N_{Am}$-Cr-X | 93.4(2) |
| $N_{Py}$-Cr-X | 92.28(7) | $N_{Py}$-Cr-X | 89.22(13) | $N_{Py}$-Cr-X | 90.7(3) |
| X-Cr-X | 133.68(3) | X-Cr-X | 142.54(4) | X-Cr-X | 140.26(7) |

$N_{Am} = N_{Amido}$, $N_{Py} = N_{Pyridine}$

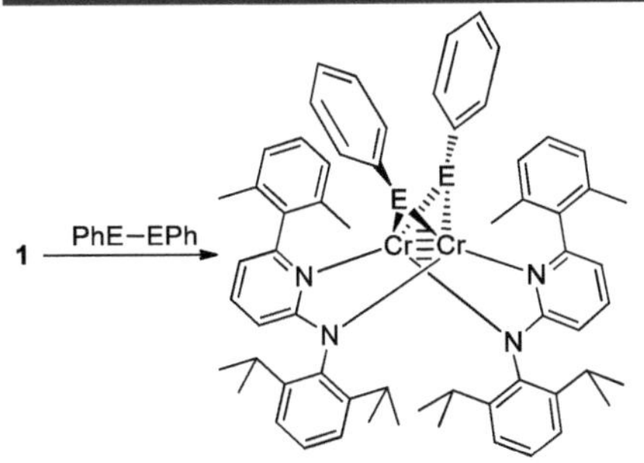

# 5. Quintuple Bond Reactivity towards Group 16 and 17 Elements; Addition vs. Insertion

**Scheme 7:** Synthesis of **9** and **10**.

The reaction of diphenyldiselenide and diphenyldisulfide with **1** at room temperature in toluene resulted in the formation of complexes **9** and **10**, respectively (Scheme 7). Complexes **9** and **10** show very little solubility in organic solvents once precipitated which excludes detailed characterization by NMR spectroscopy. Crystals of **9** suitable for X-ray analysis were grown from THF-$d_8$ solution and of **10** from toluene solution. **9** and **10** are isostructural dinuclear complexes in which the Cr-Cr bond is inserted into the S-S and Se-Se bonds, respectively (Figure 4). The two Cr atoms are joined by two bridging phenylthio or phenylselenato groups. The average Cr-S distance in **9** is 2.44 Å and in **10** the Cr-Se bond distance lies at 2.570 Å. The difference between the Cr-S and Cr-Se bond distances reflects the difference in covalent radii of S and Se (Table 4). The Cr-ligand distances in **9** and **10** are almost identical to each other. The Cr–Cr distances of 1.8486(19) Å (**9**) and 1.8635(16) Å (**10**) fall into the range of 'supershort' chromium–chromium quadruple bonds (Cr–Cr 2.0 Å).[26] Details of the X-ray crystal structure analyses are summarized in Table 3.

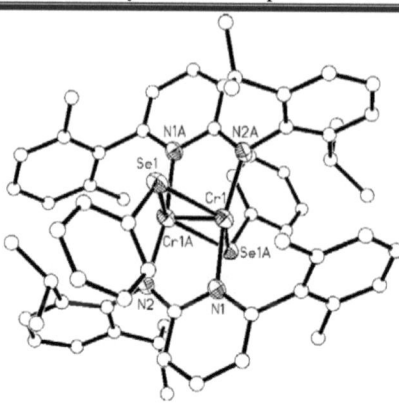

**Figure 4:** Molecular structure of **10** ORTEP representation on the 50% probability level for all non-carbon atoms. Hydrogen atoms have been deleted for clarity.

**Table 3:** Selected bond lengths [Å] and angles [°] for complexes **9** and **10**.

| 9 (E = S) | | 10 (E = Se) | |
|---|---|---|---|
| Cr-Cr | 1.8486(14) | Cr-Cr | 1.8635(16) |
| $N_{Am}$-Cr | 1.992(5) | $N_{Am}$-Cr | 2.012(5) |
| $N_{Py}$-Cr | 2.030(5) | $N_{Py}$-Cr | 2.030(5) |
| E-Cr | 2.444(19) | E-Cr | 2.5608(10) |
| Cr-Cr-E | 64.54(7) | Cr-Cr-E | 69.06(5) |
| Cr-E-Cr | 44.45(5) | Cr-E-Cr | 42.53(4) |
| Cr-Cr-$N_{Am}$ | 97.64(14) | Cr-Cr-$N_{Am}$ | 98.61(14) |
| Cr-Cr-$N_{Py}$ | 95.09(14) | Cr-Cr-$N_{Py}$ | 93.67(12) |
| $N_{Am}$-Cr-$N_{Py}$ | 167.27(19) | $N_{Am}$-Cr-$N_{Py}$ | 167.72(17) |
| $N_{Am}$-Cr-E | 86.47(15) | $N_{Am}$-Cr-E | 88.44(14) |
| $N_{Py}$-Cr-E | 98.28(10) | $N_{Py}$-Cr-E | 96.03(13) |
| E-Cr-E | 135.55(5) | E-Cr-E | 137.46(4) |

$N_{Am} = N_{Amido}$, $N_{Py} = N_{Pyridine}$

## 5.3 Conclusion

In summary, reactivity studies on a quintuply bonded, low valent dichromium complex have been extended to a variety of small inorganic compounds of group 16 and 17 elements. It has been shown that the dichromium platform can provide from two to eight electrons to give dinuclear complexes, in which Cr exits in +II and +V oxidation states. Preferentially, two electrons are donated. For the group 16 elements, $O_2$ leads to a dimeric $Cr^V$ specie. Other chalcogens undergo addition reactions in which an $E_2^{2-}$ moiety binds to the $Cr^{II}_2$-unit (E = S, Se and Te). For homo diatomic molecules of group 17 insertion of the quintuple bond into the corresponding X-X bond (X = Cl, Br and I) was observed. Complex **1** was also found to cleave the single bonds of dichalcogenides $Ph_2S_2$ and $Ph_2Se_2$ forming the corresponding oxidative addition products.

**Table 4:** Data of the X-ray crystal structure analyses for complexes **2**, **3**, **5** and **6**.

| Compounds | 2[11a] | 3 | 5 | 6. 2 x $C_7H_8$ |
|---|---|---|---|---|
| Formula | $C_{25}H_{29}CrN_2O_2$ | $C_{50}H_{58}Cr_2N_4S_2$ | $C_{50}H_{58}Cr_2N_4Te_2$ | $C_{64}H_{74}Cl_2Cr_2N_4$ |
| Mr g.mol$^{-1}$ | 441.50 | 883.12 | 1074.23 | 1074.17 |
| Crystal system | monoclinic | triclinic | triclinic | triclinic |
| Space group | P2(1)/c | P-1 | P1 | P -1 |
| a [Å] | 12.6130(10) | 10.342(1) | 10.3990 (6) | 10.584(6) |
| b [Å] | 14.6670(12) | 12.5140(14) | 11.4990 (7) | 11.856(6) |
| c [Å] | 13.6720(12) | 18.550(2) | 14.5830 (11) | 13.473(7) |
| α [°] |  | 99.296(9) | 107.170 (5) | 65.461(4) |
| β [°] | 116.302(6) | 104.196(8) | 90.903(5) | 88.144(4) |
| γ [°] |  | 105.977(8) | 107.541(5) | 67.808(4) |
| V [Å$^{-3}$] | 2267.4(3) | 2168.9(4) | 1578.09(18) | 1408.38(13) |
| Crystal size [mm$^3$] | 0.37 x 0.32 x 0.23 | 0.29 x 0.11 x 0.08 | 0.26 x 0.13 x 0.11 | 0.5 x 0.34 x 0.29 |
| $\rho_{calcd.}$ [g.cm$^{-3}$] | 1.293 | 1.352 | 1.130 | 1.266 |
| µ [mm$^{-1}$] (Mo-Kα) | 0.527 | 0.638 | 1.278 | 0.524 |
| T [K] | 133(2) | 133(2) | 133(2) | 133(2) |
| θ range [°] | 1.80 - 25.79 | 1.17 - 25.74 | 1.47 - 25.71 | 1.68 - 25.63 |
| No. unique refl. | 4289 | 8203 | 11305 | 5307 |
| No. of obsd. Refl. [I > 2σ (I)] | 3181 | 2474 | 6549 | 3642 |
| No. Of parameters | 273 | 524 | 535 | 325 |
| wR$_2$ (all data) | 0.1333 | 0.2318 | 0.1063 | 0.1096 |
| R value [I > 2σ (I)] | 0.0528 | 0.0893 | 0.0505 | 0.0443 |

# 5. Quintuple Bond Reactivity towards Group 16 and 17 Elements; Addition vs. Insertion

**Table 5:** Data of the X-ray crystal structure analyses for complexes **7**, **8**, **9** and **10**.

| Compounds | **7**. 2 x $C_7H_8$ | **8**. 2 x $C_6H_6$ | **9**. 2 x $C_4D_8O$ | **10** |
|---|---|---|---|---|
| Formula | $C_{64}H_{74}Cr_2Br_2N_4$ | $C_{62}H_{70}Cr_2I_2N_4$ | $C_{70}H_{84}Cr_2N_4O_2S_2$ | $C_{62}H_{68}Cr_2N_4Se_2$ |
| Mr g.mol$^{-1}$ | 1163.09 | 1229.02 | 1181.55 | 1131.12 |
| Crystal system | triclinic | triclinic | triclinic | triclinic |
| Space group | P -1 | P -1 | P -1 | P-1 |
| a [Å] | 10.474(8) | 13.6265(13) | 11.5330(8) | 11.3100(8) |
| b [Å] | 11.937(9) | 13.6621(12) | 12.2980(10) | 11.4090(9) |
| c [Å] | 13.549(9) | 16.0303(13) | 12.9370(10) | 11.8050(9) |
| α [°] | 65.290(5) | 97.156(7) | 102.624(6) | 100.458(6) |
| β [°] | 88.590(6) | 97.922(7) | 107.411(6) | 98.588(6) |
| γ [°] | 68.247(5) | 110.392(7) | 110.733(6) | 111.989(6) |
| V [Å$^{-3}$] | 1412.16(18) | 2722.3(4) | 1525.4(2) | 1348.98(18) |
| Crystal size [mm$^3$] | 0.27 x 0.12 x 0.1 | 0.16 x 0.16 x 0.14 | 0.29 x 0.24 x 0.07 | 0.20 x 0.17 x 0.13 |
| ρ$_{calcd.}$ [g.cm$^{-3}$] | 1.368 | 1.499 | 1.286 | 1.392 |
| μ [mm$^{-1}$] (Mo-Kα) | 1.843 | 1.576 | 0.474 | 1.795 |
| T [K] | 133(2) | 133(2) | 133(2) | 133(2) |
| θ range [°] | 1.67 - 25.68 | 1.31 - 26.17 | 1.77 - 24.59 | 1.81 - 25.81 |
| No. unique refl. | 5328 | 10275 | 5105 | 4518 |
| No. of obsd. Refl.[I > 2σ (I)] | 3341 | 5433 | 2110 | 3012 |
| No. Of parameters | 328 | 635 | 367 | 322 |
| wR$_2$ (all data) | 0.1754 | 0.1961 | 0.1672 | 0.1341 |
| R value [I > 2σ (I)] | 0.0641 | 0.0609 | 0.0680 | 0.0569 |

## 5.4 Experimental Section

**General Considerations**

All manipulations were performed with rigorous exclusion of oxygen and moisture in Schlenk-type glassware on a dual manifold Schlenk line or in a $N_2$ filled glove box (mBraun 120-G) with a high-capacity recirculator (<0.1ppm $O_2$). Solvents were dried by distillation from sodium wire / benzophenone. Complex **1** was prepared according to the published procedure.[11a] Commercial $CrCl_2$, $S_8$, $Se_8$, Te, PhSSPh and PhSeSePh were used as received. Deuterated solvents were obtained from Cambridge Isotope Laboratories and were degassed, dried and distilled prior to use. NMR spectra were recorded on Brucker 250 MHz, Varian 300 MHz and Varian 400 MHz spectrometers at ambient temperature. The chemical shifts are reported in ppm relative to the internal TMS. Elemental analyses (CHN) were determined using a Vario EL III instrument. X-ray crystal structure analyses were performed by using a STOE-IPDS II equipped with an Oxford Cryostream low-temperature unit. Structure solution and refinement was accomplished using SIR97,[27] SHELXL97[28] and WinGX.[29] Crystallographic details are summarized in Table 4 and Table 5. X-ray crystallographic data for compounds **2** (CCDC 713546), **3** (CCDC 901063), **5** (CCDC 901062), **6** (CCDC 901066), **7** (CCDC 901061), **8** (CCDC 901060), **9** (CCDC 901064), and **10** (CCDC 901065). This material is available free of charge via the Internet at http://pubs.acs.org.

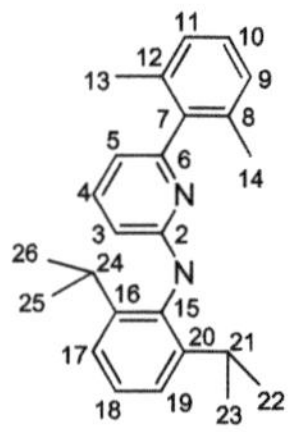

**Figure 5:** Labeling of the NMR signals.

## 5. Quintuple Bond Reactivity towards Group 16 and 17 Elements; Addition vs. Insertion

**Synthesis of 2:**[11a] $O_2$ (1 bar) was introduced into a solution of **1** (0.1 g, 0.12 mmol) in hexane (10 mL) at room temperature with continuous stirring. A colour change from purple to brown was observed within five minutes. The reaction solution was further stirred at ≈ 300 rpm for four hours and then filtered using glass fibre filters. The filtrate was kept at -30 °C to afford brown crystals overnight. Yield: 0.079 g (73%). $C_{50}H_{58}Cr_2N_4O_4$ (883.00): calcd. C 68.01 H 6.62 N 6.34; found C 67.71, H 6.88, N 6.10. $^1$H NMR (400 MHz, $C_6D_6$): $\delta$ = 1.40 (d, 12H, $J$ = 6.8 Hz, $H^{22,23,25,26}$), 1.41 (d, 12H, $J$ = 6.8 Hz, $H^{22,23,25,26}$), 1.80 (s, 6H, $H^{13,14}$), 2.24 (s, 6H, $H^{13,14}$), 2.46 (sept, 2H, $J$ = 6.8 Hz, $H^{21,24}$), 4.76 (sept, 2H, $J$ = 6.8 Hz, $H^{21,24}$), 5.40 (d, 2H, $J$ = 8.4 Hz, $H^3$), 5.85 (d, 2H, $J$ = 6.8 Hz, $H^5$), 6.74-7.38 (m, 14H, $H^{4,9,10,11,17,18,19}$) ppm. $^{13}$C NMR (100 MHz, $C_6D_6$): $\delta$ = 19.8 ($C^{13,14}$), 20.2 ($C^{13,14}$), 24.4 ($C^{22,23,25,26}$), 24.7 ($C^{22,23,25,26}$), 25.0 ($C^{22,23,25,26}$), 28.1 ($C^{21,24}$), 28.7 ($C^{21,24}$), 101.0 ($C^3$), 111.8 ($C^5$), 123.9 ($C^{9,11}$), 124.3 ($C^{9,11}$), 127.2 ($C^{18}$), 127.5 ($C^{17,19}$), 136.2 ($C^{10}$), 136.5 ($C^4$), 136.7 ($C^{8,12}$), 140.2 ($C^7$), 142.6 ($C^{8,12}$), 145.3 ($C^{16,20}$), 147.4 ($C^{15}$), 157.9 ($C^6$), 173.5 ($C^2$) ppm.

**Synthesis of 3:** Toluene (5 mL) was added to a mixture of **1** (82 mg, 0.1 mmol) and S (7 mg, 0.2 mmol) at room temperature. The reaction mixture was stirred at ≈ 300 rpm for two hours at room temperature as colour changed to brown green. Red needles were obtained by cooling a hot toluene solution to room temperature overnight. Yield 0.052 g (58%). $C_{50}H_{58}Cr_2N_4S_2$ (882.29): calcd. C 68.00, H 6.62, N 6.34; found C 67.21, H 6.23, N 6.17. $^1$H NMR (250 MHz, $C_6D_6$): $\delta$ = 0.31 (d, 6H, $J$ = 6.7 Hz, $H^{22/23/25/26}$), 1.06 (d, 6H, $J$ = 6.7 Hz, $H^{22/23/25/26}$), 1.11 (s, 6H, $H^{13,14}$), 1.14 (d, 6H, $J$ = 6.7 Hz, $H^{22/23/25/26}$), 1.38 (d, 6H, $J$ = 6.7 Hz, $H^{22/23/25/26}$), 2.24 (s, 6H, $H^{13,14}$), 2.84 (sept, 2H, $J$ = 6.7 Hz, $H^{21/24}$), 3.51 (sept, 2H, $J$ = 6.7 Hz, $H^{21/24}$), 5.73 (d, 2H, $J$ = 6.5 Hz, $H^3$), 6.27 (d, 2H, $J$ = 6.5 Hz, $H^5$), 6.33 (d, 2H, $J$ = 8.9 Hz, $H^{17/19}$), 6.64-6.79 (m, 5H, $H^{4,9,11/17,19}$), 7.00-7.05 (m, 4H, $H^{10/18}$), 7.19 (m, 2H, $H^{9/11/17/19}$) ppm. $^{13}$C NMR (75 MHz, $C_6D_6$): $\delta$ = 18.9 ($C^{13,14}$), 22.1 ($C^{13,14}$), 23.6 ($C^{22/23/25/26}$), 24.7 ($C^{22/23/25/26}$), 24.9 ($C^{22,23,/25,26}$), 25.4 ($C^{22,23,25,26}$), 27.9 ($C^{21,24}$), 29.6

($C^{21,24}$), 108.9 ($C^3$), 109.8 ($C^5$), 124.0 ($C^{10}$), 125.7 ($C^{17,19}$), 125.9 ($C^{17,19}$), 127.2 ($C^{18}$), 127.8 ($C^{18}$), 128.5 ($C^4$), 135.5 ($C^{9,11}$), 135.8 ($C^{8,12}$), 137.1 ($C^4$), 138.5 ($C^7$), 143.9 ($C^{16,20}$), 144.4 ($C^{16,20}$), 144.9 ($C^{15}$), 157.8 ($C^6$), 171.9 ($C^2$) ppm.

**Synthesis of 4:** Toluene (5 mL) was added to a mixture **1** (246 mg, 0.3 mmol) and Se (48 mg, 0.6 mmol) at room temperature. The reaction mixture was stirred at ≈ 300 rpm at 60 °C for two hours as colour changed to brown green. Red needles were obtained by cooling a hot toluene solution to room temperature overnight. Yield 0.180 g (61%) $C_{50}H_{58}Cr_2N_4Se_2$ (978.18): calcd. C 61.47, H 5.98, N 5.73; found C 61.48, H 5.76, N 5.72. $^1$H NMR (300 MHz, $C_6D_6$): $\delta$ = 0.39 (d, 6H, $J$ = 6.8 Hz, $H^{22/23/25/26}$), 1.03-1.06 (m, 12H, $H^{13,14,22,23,25,26}$), 1.11 (d, 6H, $J$ = 6.8 Hz, $H^{22/23/25/26}$), 1.39 (d, 6H, $J$ = 6.8 Hz, $H^{22/23/25/26}$), 2.18 (s, 6H, $H^{13,14}$), 3.09 (sept, 2H, $J$ = 6.8 Hz, $H^{21/24}$), 3.53 (sept, 2H, $J$ = 6.8 Hz, $H^{21/24}$), 5.74 (d, 2H, $J$ = 6.7 Hz, $H^3$), 6.27 (d, 2H, $J$ = 6.7 Hz, $H^5$), 6.32 (d, 2H, $J$ = 9.1 Hz, $H^{17/19}$), 6.64-6.77 (m, 5H, $H^{4,9,11/17,19}$), 7.02-7.05 (m, 2H, $H^{10/18}$), 7.19 (m, 4H, $H^{9/11/17/19}$) ppm. $^{13}$C NMR (75 MHz, $C_6D_6$): $\delta$ = 18.8 ($C^{13,14}$), 22.6 ($C^{13,14}$), 23.7 ($C^{22/23/25/26}$), 24.9 ($C^{22/23/25/26}$), 25.1 ($C^{22,23,/25,26}$), 25.3 ($C^{22,23/25,26}$), 27.9 ($C^{21,24}$), 29.6 ($C^{21,24}$), 109.1 ($C^3$), 110.0 ($C^5$), 124.2 ($C^{10}$), 125.7 ($C^{17,19}$), 126.1 ($C^{17,19}$), 127.1 ($C^{18}$), 128.5 ($C^4$), 134.0 ($C^{10}$), 135.5 ($C^{9,11}$), 135.2 ($C^{8,12}$), 135.7 ($C^{8,12}$), 137.3 ($C^4$), 138.6 ($C^7$), 144.3 ($C^{16,20}$), 144.4 ($C^{16,20}$), 145.2 ($C^{15}$), 157.8 ($C^6$), 172.6 ($C^2$) ppm.

**Synthesis of 5:** Toluene (5 mL) was added to a mixture of **1** (41 mg, 0.05 mmol) and Te (25 mg, 0.2 mmol) at room temperature and the reaction mixture was then heated at 60 °C and stirred at ≈ 300 rpm for three hours as colour changed to brown green. The solution was filtered using glass fibre filters and the volume of the filtrate was reduced to ca. 1 mL to afford red crystals overnight at low temperature. Yield 0.025 g (47%). $C_{50}H_{58}Cr_2N_4Te_2 \cdot C_7H_8$ (1074.21): calcd: C 58.70, H 5.70, N 4.80; found C 58.46, H 6.09, N 5.35. $^1$H NMR (250 MHz, $C_6D_6$): $\delta$ = 0.49 (d, 6H, $J$ = 6.7 Hz, $H^{22/23/25/26}$), 0.96 (s, 6H,

# 5. Quintuple Bond Reactivity towards Group 16 and 17 Elements; Addition vs. Insertion

$H^{13,14}$), 1.01-1.06 (m, 12H, $H^{22/23/25/26}$), 1.42 (d, 6H, $J$ = 6.7 Hz, $H^{22/23/25/26}$), 2.11 (s, 12H, $H^{13,14}$), 3.47 (sept, 2H, $J$ = 6.7 Hz, $H^{21/24}$), 3.59 (sept, 2H, $J$ = 6.7 Hz, $H^{21/24}$), 5.79 (d, 2H, $J$ = 4.9 Hz, $H^3$), 6.25 (d, 2H, $J$ = 6.5 Hz, $H^5$), 6.33 (d, 2H, $J$ = 8.9 Hz, $H^{17/19}$), 6.66-6.77 (m, 5H, $H^{4/10}$), 7.03-7.18 (m, 6H, $H^{10/18, 9/11/17/19}$) ppm.

**Synthesis of 6:** $Cl_2$ (ca. 0.2 bar) was introduced into a solution of **1** (0.4 g, 0.48 mmol) in toluene (10 mL) at room temperature with continuous stirring. A colour change from purple to red-brown was observed within 10 minutes. The mixture was stirred at ≈ 300 rpm under $Cl_2$ atmosphere overnight after which some dark precipitate had formed. Then the solution was filtered using glass fibre filters and the filtrate was kept at -30°C to afford brown crystals. Yield 0.032g (33%). $C_{50}H_{58}Cr_2N_4Cl_2 \cdot C_7H_8$ (980.3): calcd. C 69.71, H 6.77, N 5.71; found C 69.95, H 6.81, N 5.26. $^1$H NMR (400 MHz, $C_6D_6$, 298 K): $\delta$ = 1.09 (d, 12H, $J$ = 7.5 Hz, $H^{22/23/25/26}$), 1.40 (d, 12H, $J$ = 7.5 Hz, $H^{22/23/25/26}$), 2.80 (s, 6H, $H^{13/14}$), 4.71 (sept, 2H, $J$ = 7.5 Hz, $H^{21/24}$), 5.20 (d, 2H, $J$ = 7.5 Hz, $H^3$), 6.17 (d, 2H, $J$ = 5.2 Hz, $H^5$), 6.38 (t, 2H, $J$ = 5.2 Hz, $H^4$), 6.98-7.37 (m, 12H, $H^{9/10/11, 17/18/19}$) ppm. $^{13}$C NMR ($C_6D_6$, 298 K): $\delta$ = 20.5 ($C^{13/14}$), 23.8 ($C^{22/23/25/26}$), 28.7 ($C^{21/24}$), 103.7 ($C^3$), 114.5 ($C^5$), 124.2 ($C^{9/11}$), 125.6 ($C^{17/19}$), 127.7 ($C^{18}$), 128.0 ($C^{10}$), 134.1 ($C^{8/12}$), 135.9 ($C^{8/12}$), 137.9 ($C^4$), 142.0 ($C^7$), 148.1 ($C^{16/20}$), 154.3 ($C^{15}$), 159.5($C^6$), 160.0 ($C^2$) ppm.

**Synthesis of 7:** $Br_2$ (7 μL, 0.1 mmol) was added to a solution of **1** (82 mg, 0.1 mmol) in toluene (10 mL) at room temperature with continuous stirring at ≈ 300 rpm. A colour change from purple to brown was observed. The reaction solution was filtered using glass fibre filters and the filtrate was kept at -30°C to afford brown crystals. Yield 0.052g (48%). $C_{50}H_{58}Cr_2N_4Br_2 \cdot C_7H_8$ (1070.7): calcd. C 63.92, H 6.21, N 5.23; found C 64.17, H 5.39, N 5.35. $^1$H NMR (400 MHz, $C_6D_6$, 298 K): $\delta$ = 1.10 (d, 12H, $J$ = 6.5 Hz $H^{22/23/25/26}$), 1.32 (d, 12H, $J$ = 6.5 Hz, $H^{22/23/25/26}$), 2.38 (s, 6H, $H^{13/14}$), 5.37 (sept, 2H, $J$ = 6.8 Hz, $H^{21/24}$), 5.52 (d, 2H, $J$ = 6.8 Hz, $H^3$), 6.15 (d, 2H, $J$ = 6.5 Hz, $H^5$), 6.28 (t, 2H, $J$ =

6.5 Hz, H$^4$), 6.49 (m, 6H, H$^{17/18/19}$), 6.96-7.36 (m, 12H, H$^{9/10/11}$) ppm. $^{13}$C NMR (C$_6$D$_6$, 298 K): $\delta$ = 19.8 (C$^{13/14}$), 20.5 (C$^{13/14}$), 23.7 (C$^{22/23/25/26}$), 24.6 (C$^{22/23/25/26}$), 25.7 (C$^{22/23/25/26}$), 28.7 (C$^{21/24}$), 103.7 (C$^3$), 114.6 (C$^5$), 124.2 (C$^{9/11}$), 124.9 (C$^{17/19}$), 125.6 (C$^{18}$), 127.7 (C$^{10}$), 130.1 (C$^{8/12}$), 135.8 (C$^{8/12}$), 137.2 (C$^4$), 137.9 (C$^7$), 142.2 (C$^{16/20}$), 148.0 (C$^{15}$), 148.7 (C$^6$), 160.4 (C$^2$) ppm.

**Synthesis of 8:** I$_2$ (6 mg, 0.025 mmol) was added to a solution of **1** (21 mg, 0.025 mmol) in C$_6$D$_6$ (0.5 mL) at room temperature and was shaken briefly. A colour change from purple to brown was observed. Standing of the solution at room temperature afforded red crystals. C$_{50}$H$_{58}$Cr$_2$N$_4$I$_2$ (1072.16): Calc. C 55.98, H 5.45, N 5.22; found. C 55.99, H 5.88, N 4.78. $^1$H NMR (400 MHz, C$_6$D$_6$): $\delta$ = 1.14 (br d, 12H, H$^{22,23,25,26}$), 1.24 (br d, 12H, H$^{22,23,25,26}$), 2.19 (s, 6H, H$^{13,14}$), 3.97 (br sept, 2H, H$^{21,24}$), 5.74 (d, 2H, $J$ = 6.8 Hz, H$^3$), 6.15 (d, 2H, $J$ = 6.5 Hz, H$^5$), 6.28 (t, 2H, $J$ = 6.5 Hz, H$^4$), 6.49 (m, 6H, H$^{17,18,19}$), 6.96-7.36 (m, 12H, H$^{9,10,11}$) ppm. $^{13}$C NMR (C$_6$D$_6$, 298 K): $\delta$ = 19.8 (C$^{13,14}$), 20.5 (C$^{13,14}$), 23.7 (C$^{22,23,25,26}$), 24.6 (C$^{22,23,25,26}$), 25.7 (C$^{22,23,25,26}$), 28.7 (C$^{21,24}$), 103.7 (C$^3$), 108.1 (C$^5$), 124.2 (C$^{9,11}$), 124.9 (C$^{17,19}$), 125.6 (C$^{18}$), 127.7 (C$^{10}$), 134.7 (C$^{8,12}$), 135.8 (C$^{8,12}$), 137.2 (C$^4$), 140.2 (C$^7$), 142.2 (C$^{16,20}$), 148.0 (C$^{15}$), 148.7 (C$^6$), 160.0 (C$^2$) ppm.

**Synthesis of 9:** Toluene (20 mL) was added to a mixture of **1** (82 mg, 0.1 mmol) and Ph$_2$SSPh$_2$ (22 mg, 0.1 mmol) at room temperature and the resulting green reaction mixture was stirred at $\approx$ 300 rpm at room temperature for six hours. Standing of the solution at room temperature afforded red crystalline material over 48 hours. Yield 0.072 g (69%). C$_{62}$H$_{68}$Cr$_2$N$_4$S$_2$ (1037.36): calcd. C 71.79, H 6.61, N 5.40; found C 71.94, H 6.60, N 5.54. $^1$H NMR (300 MHz, C$_6$D$_6$, 298 K): $\delta$ = 0.87 (br d, 12H, H$^{22/23/25/26}$), 0.96 (br d, 12H, H$^{22/23/25/26}$), 2.19 (s, 12H, H$^{13/14}$), 3.80 (br sept, 4H, H$^{21/24}$), 5.95 (d, 2H, $J$ = 6.3 Hz, H$^3$), 6.42 (br t, 2H, H$^4$), 6.49 (d, 2H, $J$ = 8.5 Hz, H$^5$), 6.90-7.41 (m, 16H, H$^{9/10/11,17/18/19/Ph}$) ppm.

**Synthesis of 10:** Toluene (20 mL) was added to a mixture of **1** (82 mg, 0.1 mmol) and Ph$_2$SeSePh$_2$ (32 mg, 0.1 mmol) at room temperature and the resulting green reaction mixture was stirred at ≈ 300 rpm at room temperature for six hours. Standing of the solution at room temperature afforded red crystalline material overnight. Yield 0.083 g (73%). C$_{62}$H$_{68}$Cr$_2$N$_4$Se$_2$ (1131.14): calcd. C 65.83, H 6.06, N 4.95; found C 64.65, H 5.90, N 4.77. $^1$H NMR (300 MHz, C$_6$D$_6$, 298 K): $\delta$ = 0.47 (m, 12H, H$^{22/23/25/26}$), 1.35 (br d, 6H, H$^{22/23/25/26}$), 1.42 (br d, 6H, H$^{22/23/25/26}$), 1.96 (s, 6H, H$^{13/14}$), 2.40 (s, 6H, H$^{13/14}$), 2.96 (sept, 2H, $J$ = 7.5 Hz, H$^{21/24}$), 4.71 (sept, 2H, $J$ = 7.5 Hz, H$^{21/24}$), 5.95 (d, 2H, $J$ = 7.5 Hz, H$^3$), 6.42 (t, 2H, $J$ = 6.9 Hz, H$^4$), 6.52 (d, 2H, $J$ = 8.7 Hz, H$^5$), 6.65-7.05 (m, 16H, H$^{9/10/11,17/18/19/\text{Ph}}$) ppm.

**Acknowledgements**

Financial support from the Deutsche Forschungsgemeinschaft (DFG KE 756/20-1) is gratefully acknowledged. E.S.T. thanks the Deutscher Akademischer Austausch Dienst (DAAD) for a Ph.D. scholarship.

## 5.5 References

[1] a) F. Wagner, A. Noor, R. Kempe, *Nat. Chem.* **2009**, *1*, 529–536; b) F. A. Cotton, L. A. Murillo, R. A. Walton, *Multiple Bonds Between Metal Atoms*, 3rd ed., Springer: Berlin, **2005**.

[2] B. O. Roos, A. C. Borin, L. Gagliardi, *Angew. Chem.* **2007**, *119*, 1491-1494; *Angew. Chem. Int. Ed.* **2007**, *46*, 1469-1472.

[3] M. D. Morse, *Chem. Rev.* **1986**, *86*, 1049-1109.

[4] a) Y. M. Efremov, A. N. Samoilova, L. V. Gurvich, *Opt. Spektrosc.* **1974**, *36*, 654-657; b) E. P. Künding, M. Moskovits, G. A. Ozin, *Nature*, **1975**, *254*, 503-504; c) W.

Klotzbücher, G. A. Ozin, *Inorg. Chem.* **1977**, *16*, 984-987; d) V. E. Bondybey, J. H. English, *Chem. Phys. Let.* **1983**, *94*, 443-447.

[5] T. Nguyen, A. D. Sutton, M. Brynda, J. C. Fettinger, G. J. Long, P. P. Power, *Science* **2005**, *310*, 844-847.

[6] a) K. A. Kreisel, G. P. A. Yap, O. Dmitrenko, C. R. Landis, K. H. Theopold, *J. Am. Chem. Soc.* **2007**, *129*, 14162–14163; b) R. Wolf, C. Ni, T. Nguyen, M. Brynda, G. J. Long, A. D. Fischer, R. C. Sutton, J. C. Fettinger, M. Hellman, L. Pu, P. P. Power, *Inorg. Chem.* **2007**, *46*, 11277–11290; c) A. Noor, F. R. Wagner, R. Kempe, *Angew. Chem.* **2008**, *120*, 7356–7359; *Angew. Chem. Int. Ed.* **2008**, *47*, 7246–7249; d) Y.-C. Tsai, C.-W. Hsu, J.-S. K. Yu, G.-H. Lee, Y. Wang, T.-S. Kuo, *Angew. Chem.* **2008**, *120*, 7360–7363; *Angew. Chem. Int. Ed.* **2008**, *47,* 7250–7253; e) C.-W. Hsu, J.-S. K. Yu, C.-H. Yen, G.-H. Lee, Y. Wang, Y.-C. Tsai, *Angew. Chem.* **2008**, *120,* 10081–10084; *Angew. Chem. Int. Ed.* **2008**, *47*, 9933–9936.

[7] A. Noor, R. Kempe, *Chem. Record* **2010**, *10*, 413-416.

[8] A. Noor, G. Glatz, R. Müller, M. Kaupp, S. Demeshko, R. Kempe, *Z. Anorg. Allg. Chem.* **2009**, *635*, 1149–1152.

[9] Y.-L. Huang, D.-Y. Lu, H.-C. Yu, J.-S. K. Yu, C.-W. Hsu, T.-S. Kuo, G.-H. Lee, Y. Wang, Y.-C. Tsai, *Angew. Chem.* **2012**, *124*, 7901–7905; *Angew. Chem. Int. Ed.* **2012**, *51*, 7781–7785.

[10] a) Y.-C. Tsai, H.-Z. Chen, C.-C. Chang, J.-S. K. Yu, G.-H. Lee, Y. Wang, T.-S. Kuo, *J. Am. Chem. Soc.* **2009**, *131*, 12534–12535; b) S.-C. Liu, W.-L. Ke, J.-S. K. Yu, T.-S. Kuo, Y.-C. Tsai, *Angew. Chem.* **2012**, *124*, 6500–6503; *Angew. Chem. Int. Ed.* **2012**, *51*, 6394–6397.

[11] a) A. Noor, G. Glatz, R. Müller, M. Kaupp, S. Demeshko, R. Kempe, *Nat. Chem.* **2009**, *1*, 322–325; b) A. Noor, E. S. Tamne, S. Qayyum, T. Bauer, R. Kempe, *Chem. Eur. J.* **2011**, *17*, 6900–6903; c) C. Schwarzmaier, A. Noor, G. Glatz, M. Zabel, A. Y.

Timoshkin, B. M. Cossairt, C. C. Cummins, R. Kempe, M. Scheer, *Angew. Chem.* **2011**, *123*, 7421–7424; *Angew. Chem. Int. Ed.* **2011**, *50*, 7283–7286.

[12] C. Ni, B. D. Ellis, G. J. Long, P. P. Power, *Chem. Commun.* **2009**, 2332–2334.

[13] J. Shen, G. P. A. Yap, J.-P. Werner, K. H. Theopold, *Chem. Commun.* **2011**, *47*, 12191–12193.

[14] H.-Z. Chen, S-C. Liu, C-H. Yen, J.-S. K. Yu, Y-J. Shieh, T.-S. Kuo, Y.-C. Tsai, *Angew. Chem.* **2012**, DOI: 10.1002/anie.201205027; *Angew. Chem. Int. Ed.* **2012**, DOI: 10.1002/anie.201205027.

[15] a) H. Nishino, J. K. Kochi, *Inorg. Chim. Acta* **1990**, *174*, 93–102; b) A. A. Danopoulos, G. Wilkinson, T. K. N. Sweet, M. B. Hursthouse, *Polyhedron* **1996**, *15*, 873–879.

[16] M. Herberhold, W. Kremnitz, A. Razavi, H. Schollhorn, U. Thewalt, *Angew. Chem.* **1985**, *97*, 603–604; *Angew. Chem. Int. Ed.* **1985**, *24*, 601–602.

[17] A. Muller, W. Jaegermann, *Inorg. Chem.* **1979**, *18*, 2631–2632.

[18] L. Y. Goh, T. C. W. Mak, *J. Chem. Soc., Chem. Commun.* **1986**, 1474–1476.

[19] H. Brunner, J. Pfauntsch, J. Wachter, B. Nuber, M. L. Ziegler, *J. Organomet Chem.*, **1989**, *359*, 179–188.

[20] H. Brunner, J. Wachter, E. Guggolz, M. L. Ziegler, *J. Am. Chem. Soc.* **1982**, *104*, 1765–1766.

[21] F. Preuss, M. Billen, F. Tabellion, G. Wolmershäuser, *Z. Anorg. Allg. Chem.*, **2000**, *626*, 2446–2448.

[22] R. E. Bachman, K. H. J. Whitmire, *Organomet. Chem.* **1994**, *479*, 31–35.

[23] C. Ni, P. P. Power, *Structure and Bonding,* Berlin, **2010**, vol 136 (Metal-Metal Bonding), pp.59–112.

[24] a) L. B. Handy, J. K. Ruff, L. F. Dahl, *J. Am. Chem. Soc.* **1970**, *92*, 7327–7337; b) D. B. Morse, T. B. Rauchfuss, S. R. Wilson, *J. Am. Chem. Soc.* **1990**, *112*, 1860–1864;

c) M. E. Burin, M. V. Smirnova, G. K. Fukin, E. V. Baranov, M. N. Bochkarev, *Eur. J. Inorg. Chem.* **2006**, 351–356.

[25] S. Deeken, G. Motz, R. Kempe, *Z. Anorg. Allg. Chem.* **2007**, *633*, 320–325.

[26] F. A. Cotton, *Multiple Bonds Between Metal Atoms*, Wiley, New York, **1982**.

[27] A. Altomare, M. C. Burla, M. Camalli, G. L. Cascarano, C. Giacovazzo, A. Guagliardi, A. G. G. Moliterni, G. Polidori, R. Spagna, *J. Appl. Cryst.* **1999**, *32*, 115–119.

[28] G. M. Sheldrick, *Acta Cryst. A* **2008**, *64*, 112-122.

[29] L. J. Farrugia, *J. Appl. Cryst.* **1999**, *32*, 837–838.

# 6. Reaction of a Cr-Cr Quintuple Bond with Phosphine Ligands

Emmanuel Sobgwi Tamne, Awal Noor, Tobias Bauer, and Rhett Kempe*[a]

[a] Inorganic Chemistry II, University of Bayreuth, 95440 Bayreuth, Germany.
E-mail: kempe@uni-bayreuth.de

**Keywords**: Chromium, metal-metal bond, N- ligands, oxidative addition, quintuple bond.

**To be submitted to:** *Z. Anorg. Allg. Chem.*

**Abstract.** The reactivity of the formally quintuply bonded bimetallic aminopyridinato chromium complex **1** towards various phosphines was investigated. Oxidative additions were observed in which the Cr-Cr quintuple bond is inserted into the corresponding P-X bond of the used phosphines $R_2PX$ (**2a**: R = $C_6H_5$, X = H; **2b**: R = $C_6H_5$, X = Cl; **2c**: R = $C_2H_5$, X = Cl). The synthesized compounds are diamagnetic; **2a** and **2c** have additionally been characterized by X-ray crystal structure analysis. The solid-state structures indicate a reduction of the bond order of the $Cr_2$ moiety to four. The bimetallic $Cr^{II}$ complexes feature metal-metal distances of 1.856(11) and 1.867(7) Å, respectively.

## 6.1 Introduction

Compounds with bond orders higher than three have been investigated by natural scientists and especially chemists for nearly 50 years.[1] Molecules with a very high bond order, namely sextuply bonded molecules, have a similarly long history. The $M_2$ molecules of the group 6 metals (M = Cr, Mo, W) were stabilized in an inert matrix or in the gas phase as short-lived species.[2] Unfortunately, the lack of a selective bulk synthesis has restricted the application of these molecules in inorganic synthesis. The

# 6. Reaction of a Cr-Cr Quintuple Bond with Phosphine Ligands

large number of electrons localized between the two metals could enable unique reactivity. In 2005, the group of Power reported a breakthrough in this regard. They synthesized the first stable molecule with a Cr-Cr quintuple bond.[3] A few years later the groups of Theopold,[4] Tsai[5] and us[6] synthesized related N-ligand stabilized coordination compounds with such high bond orders and very short metal-metal bonds. Moreover, the isolation of quintuply bonded molybdenum compounds was reported.[7] All these quintuply bonded metal complexes have been involved in the activation of small molecules. For instance the carboalumination,[8a] activation of $N_2O$ and $RN_3$ molecules,[8b] cycloaddition of alkynes and dienes,[8c,8d,8e] and the addition reactions of group 15 elements[8f] were reported. Here, the reaction of phosphine ligands towards the Cr-Cr quintuple bond is discussed.

## 6.2 Results and Discussion

We have explored the reaction behaviour of quintuply bonded chromium complex **1**[8a] towards various phosphine ligands of the general formula $R_2PX$ (R = $C_6H_5$, $C_2H_5$; X = H, Cl). Reaction of a toluene solution of **1** with the corresponding phosphine at room temperature yielded selective oxidative addition products **2a**, **2b**, and **2c**, respectively (Scheme 1). The obtained complexes are diamagnetic and have been characterized by NMR and X-ray crystal structure analysis as well as by elemental analysis.

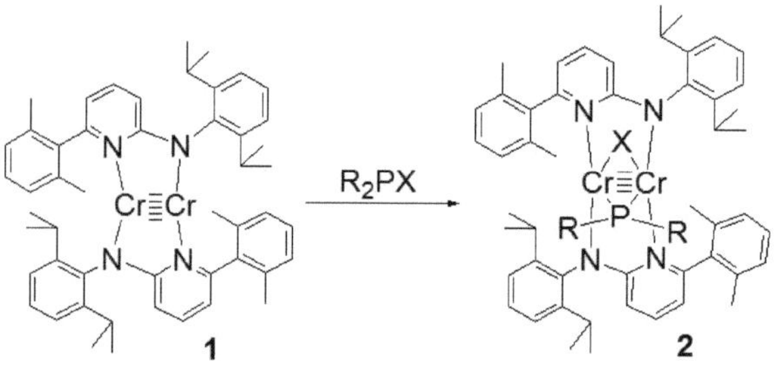

**2a**: R = R = $C_6H_5$; X = H
**2b**: R = R = $C_6H_5$; X = Cl
**2c**: R = R = $C_2H_5$; X = Cl

**Scheme 1:** Synthesis of phosphines complexes.

In the $^1$H NMR spectrum of **2a** the signal of the bridging hydride appears as a doublet at -2.05 ppm ($^2J_{P,H}$= 27 Hz). In addition, eight doublets for the non-equivalent $CH_3$-protons of the isopropyl groups, four singlets for the $CH_3$-protons and four septets corresponding to the CH-proton of the isopropyl groups were observed. Owing to the coupling of bridged phosphorus with the bridging hydride, the $^{31}$P NMR spectrum of compound **2a** shows a doublet at 80.5 ppm ($^2J_{P,H}$= 27 Hz). The complexes **2b** and **2c** show a different NMR pattern that is four doublets for the non-equivalent $CH_3$-protons of the isopropyl groups, two singlets for the $CH_3$-protons and two septets for the CH-proton of the isopropyl groups. However, a single set of signals was observed for the aromatic protons. The $^{31}$P NMR spectra of the complexes **2b** and **2c** show singlets at 58.8 and 70.8 ppm, respectively.

Crystals of **2a** suitable for X-ray analysis were grown from toluene solution and the molecular structure is presented in Figure 1. Analogous oxidative addition has been observed if **1** is reacted with diphenylchlorophosphine and diethylchlorophosphine to give **2b** and **2c**, respectively (Scheme 1). For **2c** standard workup and crystallization from hexane yielded crystals suitable for X-ray analysis (Figure 2).

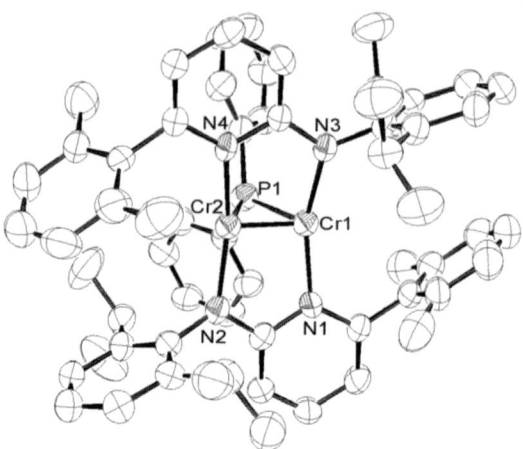

**Figure 1:** Molecular structure of 2a. [ORTEP representation (on the 50 % probability level) for all non-carbon atoms, Hydrogen atoms have been deleted for clarity.] Selected bond lengths [Å] and angles [°]: N1-Cr1 2.018(5), N2-Cr2 2.015(5), N3-Cr1 1.996(5), N4-Cr2 2.021(5), Cr1-Cr2 1.856(11), Cr1-P1 2.398(2), Cr2-P1 2.390(2); Cr1-Cr2-N2 97.4(13), Cr2-Cr1-N1 95.1(13), N2-Cr2-N4 159.6(2), N1-Cr1-N3 159.6(2), Cr1-Cr2-P1 67.42(7), Cr2-Cr1-P1 66.97(7), Cr2-P1-Cr1 45.6(4).

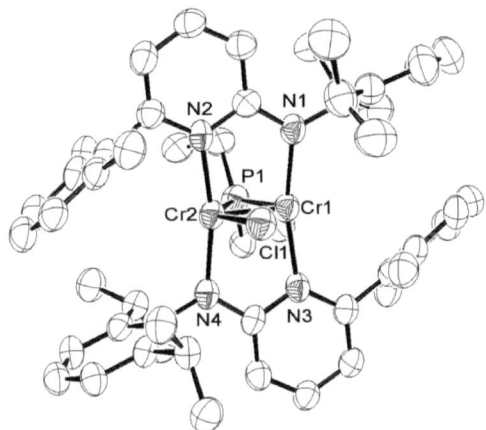

**Figure 2:** Molecular structure of **2c**. [ORTEP representation (on the 50 % probability level) for all non-carbon atoms, Hydrogen atoms have been deleted for clarity.] Selected bond lengths [Å] and angles [°]: N1-Cr1 2.002(3), N2-Cr2 2.024(3), N3-Cr1 2.041(3), N4-Cr2 1.994(3), Cr1-Cr2 1.867(7), Cr1-P1 2.365(12), Cr2-P1 2.385(11), Cr1-Cl1 2.557(11), Cr2-Cl1 2.450(11); Cr1-Cr2-N2 94.01(8), Cr2-Cr1-N1 97.98(8), Cr1-Cl1-Cr2 43.73(2), Cr1-Cr2-P1 66.23(3), Cr2-Cr1-P1 67.44(7), Cr2-P-Cr1 46.30(3).

For **2a**, the average Cr-P bond distance of 2.394 Å is slightly longer than the corresponding average bond distance of 2.375 Å for **2c**. These Cr-P bond lengths are consistent with the values observed for other phosphide-bridged Cr-Cr species.[9] The Cr1-P1-Cr2 angle of 46.3° for **2c** is larger than the corresponding angle 45.6° for **2a**. These differences between bond lengths and angles could be the result of steric effects. The average Cr-$N_{pyridine}$ bond distance of 2.019(5) Å for **2a** and 2.032(3) Å for **2c** is longer than the average Cr-$N_{amido}$ bond distance of 2.005(5) Å for **2a** and 1.998(3) Å for **2c**. This indicates the localization of the anionic charge on the amido nitrogen atom.[10] The Cr-Cr bond length of 1.856(11) Å in **2a** is shorter than that of 1.867(7) Å for **2c**, but still in the range of super short Cr-Cr quadruply bonded compounds (Cr-Cr < 2 Å).[1b]

# 6. Reaction of a Cr-Cr Quintuple Bond with Phosphine Ligands

**Table 1:** Data of the X-ray crystal structure analyses for complexes **2a** and **2c**.

| Compound | 2a | 2c |
|---|---|---|
| Formula | $C_{62}H_{69}Cr_2N_4P$ | $C_{54}H_{68}Cr_2N_4PCl$ |
| Mr g.mol$^{-1}$ | 1005.2 | 943.5 |
| Crystal system | monoclinic | monoclinic |
| Space group | $P2_1/n$ | $C2/c$ |
| a [Å] | 11.609(6) | 37.170(10) |
| b [Å] | 18.517(10) | 12.528(5) |
| c [Å] | 24.879(12) | 25.417(10) |
| α [°] | 90 | 90 |
| β [°] | 98.626(4) | 108.753(3) |
| γ [°] | 90 | 90 |
| V [Å$^{-3}$] | 5287.6(5) | 11207.5(7) |
| Crystal size [mm$^3$] | 0.31 x 0.21 x 0.12 | 0.38 x 0.32 x 0.29 |
| $\rho_{calcd.}$ [g.cm$^{-3}$] | 1.261 | 1.118 |
| µ [mm$^{-1}$] (Mo-Kα) | 0.484 | 0.499 |
| T [K] | 133(2) | 133(2) |
| θ range [°] | 1.38-25.64 | 1.16-25.67 |
| No. unique refl. | 9965 | 10590 |
| No. of obsd. Refl.[I > 2σ (I)] | 3612 | 5151 |
| No. of parameters | 634 | 573 |
| wR$_2$ (all data) | 0.1594 | 0.1147 |
| R value [I > 2σ (I)] | 0.0694 | 0.0481 |

## 6.3 Conclusion

In summary, quintuply bonded chromium complex **1** undergoes oxidative addition reactions with phosphine ligands at room temperature to give phosphorus bridged chromium complexes. For all obtained complexes, reduction in the bond order was observed that lead to products featuring a quadruply bonded Cr-Cr moiety.

## 6.4 Experimental Section

**General Procedures**

All manipulations were performed with rigorous exclusion of oxygen and moisture in Schlenk-type glassware on a dual manifold Schlenk line or in an $N_2$-filled glove box (mBraun 120-G) with a high capacity recirculator (< 0.1 ppm $O_2$). Solvents were dried by distillation from sodium wire/benzophenone. Complex **1** was prepared according to a published procedure.[8a] Commercial $CrCl_2$, $Ph_2PH$, $Ph_2PCl$ and $Et_2PCl$ (Alfa Aesar, Acros) were used as received. Deuterated solvents were obtained from Cambridge Isotope Laboratories and were degassed, dried and distilled prior to use. NMR spectra were recorded on Bruker 250 MHz, Varian 400 MHz and Varian 500 MHz instruments at ambient temperature. The chemical shifts are reported in ppm relative to the internal TMS. Elemental analyses (CHN) were determined using a Vario EL III instrument. X-ray crystal structure analyses were performed by using a STOE-IPDS II equipped with an Oxford Cryostream low-temperature unit. Structure solution and refinement was accomplished using SIR97,[11] SHELXL97[12] and WinGX.[13] CCDC-XXX for **2a** and CCDC-XXY for **2c** contain the supplementary crystallographic data for this paper. These data can be obtained free of charge at www.ccdc.cam.ac.uk/conts/retrieving.html (or from the Cambridge Crystallographic Data Centre, 12 Union Road, Cambridge CB21 EZ, UK, Fax:+44-1223-336-033, E-mail: deposit@ccdc.cam.ac.uk).

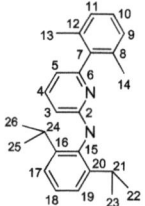

**Figure 3:** Labeling of the NMR signals.

**Synthesis of 2a:** Ph$_2$PH (17.4 μL, 0.1 mmol) was added to a solution of **1** (0.082 g, 0.1 mmol) in toluene (10 mL). A colour change from purple to brown was observed. The reaction mixture was stirred for one hour at room temperature. The mixture was then filtered and the filtrate was concentrated in vacuo till it appeared oily. This product then afforded X-ray quality crystals of **2a** upon standing overnight at room temperature. Yield 0.062 g (62 %). C$_{62}$H$_{69}$Cr$_2$N$_4$P (1005.20): calcd. C 74.08, H 6.92, N 5.57; found C 74.66, H 7.37, N 5.36. $^1$H NMR (400 MHz, C$_6$D$_6$): $\delta$ = -2.05 (d, 1H, $^2J_{PH}$ = 27.2 Hz, μ-H), 0.47 (d, 3H, $^3J_{HH}$ = 6.8 Hz, H$^{22/23/25/26}$), 0.52 (d, 3H, $^3J_{HH}$ = 6.8 Hz, H$^{22/23/25/26}$), 0.57 (d, 3H, $^3J_{HH}$ = 6.8 Hz, H$^{22/23/25/26}$), 0.88 (d, 3H, $^3J_{HH}$ = 6.8 Hz, H$^{22/23/25/26}$), 1.23 (d, 3H, $^3J_{HH}$ = 6.8 Hz, H$^{22/23/25/26}$), 1.29 (d, 3H, $^3J_{HH}$ = 6.4 Hz, H$^{22/23/25/26}$), 1.47 (d, 3H, $^3J_{HH}$ = 6.8 Hz, H$^{22/23/25/26}$), 1.49 (d, 3H, $^3J_{HH}$ = 6.8 Hz, H$^{22/23/25/26}$), 2.27 (s, 6H, H$^{13/14}$), 2.29 (s, 6H, H$^{13/14}$), 2.31 (s, 6H, H$^{13/14}$), 2.55 (s, 6H, H$^{13/14}$), 2.69 (sep, 2H, $^3J_{HH}$ = 6.8 Hz, H$^{21/24}$), 2.96 (sep, 2H, $^3J_{HH}$ = 6.8 Hz, H$^{21/24}$), 4.07 (sep, 2H, $^3J_{HH}$ = 6.8 Hz, H$^{21/24}$), 4.20(sep, 2H, $^3J_{HH}$ = 6.8 Hz, H$^{21/24}$), 5.96 (d, 2H, $^3J_{HH}$ = 6.8 Hz, H$^3$), 6.17 (d, 2H, $^3J_{HH}$ = 7.2 Hz, H$^5$), 6.25 (t, 2H, $^3J_{HH}$ = 4.4 Hz, H$^4$), 6.48 (t, 6H, $^3J_{HH}$ = 5.2 Hz, H$^{PPh}$), 6.95-7.20 (m, 12H, H$^{9/10/11,17/18/19}$), 7.35 (m, 4H, H$^{PPh}$), 7.46 (t, 6H, $^3J_{HH}$ = 9.2 Hz, H$^{PPh}$) ppm. $^{13}$C NMR (100 MHz, C$_6$D$_6$): $\delta$ = 20.4 (C$^{13/14}$), 21.6 (C$^{13/14}$), 23.6 (C$^{22/23/25/26}$), 25.1 (C$^{22/23/25/26}$), 26.8 (C$^{22/23/25/26}$), 28.0 (C$^{22/23/25/26}$), 28.7 (C$^{21/24}$), 103.7 (C$^3$), 114.5 (C$^5$), 124.1 (C$^{9/11}$), 124.5 (C$^{17/19}$), 125.4 (C$^{18}$), 127.7 (C$^{10}$), 128.2 (C$^{PPh}$), 128.6 (C$^{PPh}$), 128.8 (C$^{PPh}$), 132.2 (C$^{8/12}$), 134.2 (C$^{PPh}$), 134.7 (C$^{8/12}$), 141.8 (C$^4$), 142.1 (C$^7$), 145.4 (C$^{PPh}$), 147.9(C$^{16/20}$), 159.2 (C$^6$), 159.8(C$^2$) ppm. $^{31}$P NMR (500 MHz, C$_6$D$_6$): $\delta$ = 80.5 (d, 1P, $^2J_{PH}$ = 27.0 Hz, μ-P) ppm.

**Synthesis of 2b:** Ph$_2$PCl (18 µL, 0.1 mmol) was added to a solution of **1** (0.082 g, 0.1 mmol) in toluene (10 mL). A colour change from purple to brown was observed. The reaction mixture was stirred for two hours at room temperature. Then, the solution was filtered and the filtrate was kept at -30°C to afford red-brown crystals. Yield 0.034 g (30 %). C$_{62}$H$_{68}$Cr$_2$N$_4$PCl (1038.37): calcd. C 71.63, H 6.59, N 5.39; found C 71.20, H 6.75, N 5.19. $^1$H NMR (400 MHz, C$_6$D$_6$): $\delta$ = 0.28 (d, 12H, $^3J_{HH}$ = 6.0 Hz, H$^{22/23/25/26}$), 0.87 (d, 12H, $^3J_{HH}$ = 7.2 Hz, H$^{22/23/25/26}$), 1.23 (d, 12H, $^3J_{HH}$ = 6.4 Hz, H$^{22/23/25/26}$), 1.43 (d, 12H, $^3J_{HH}$ = 6.0 Hz, H$^{22/23/25/26}$), 2.30 (s, 6H, H$^{13/14}$), 2.37 (s, 6H, H$^{13/14}$), 2.67 (sep, 2H, $^3J_{HH}$ = 6.8 Hz, H$^{21/24}$), 4.19 (sep, 2H, $^3J_{HH}$ = 6.8 Hz, H$^{21/24}$), 5.98 (d, 2H, $^3J_{HH}$ = 6.4 Hz, H$^3$), 6.29 (d, 2H, $^3J_{HH}$ = 9.6 Hz, H$^5$), 6.41 (t, 2H, $^3J_{HH}$ = 8.0 Hz, H$^4$), 6.92 (t, 6H, $^3J_{HH}$ = 6.4 Hz, H$^{PPh}$), 7.01-7.24 (m, 12H, H$^{9/10/11,17/18/19}$), 7.34 (m, 4H, H$^{PPh}$) ppm. $^{13}$C NMR (100 MHz, C$_6$D$_6$): $\delta$ = 20.8 (C$^{13/14}$), 21.7 (C$^{13/14}$), 22.8 (C$^{22/23/25/26}$), 25.7 (C$^{22/23/25/26}$), 26.7 (C$^{22/23/25/26}$), 28.1 (C$^{22/23/25/26}$), 29.1 (C$^{21/24}$), 107.4 (C$^3$), 114.9 (C$^5$), 123.8 (C$^{9/11}$), 126.0 (C$^{17/19}$), 126.2 (C$^{18}$), 128.3 (C$^{10}$), 128.5 (C$^{PPh}$), 128.9 (C$^{PPh}$), 129.0 (C$^{PPh}$), 129.6 (C$^{8/12}$), 133.8 (C$^{PPh}$), 136.9 (C$^{8/12}$), 138.4 (C$^4$), 141.2 (C$^7$), 147.2 (C$^{PPh}$), 147.7 (C$^{16,20}$), 148.4 (C$^6$), 158.5 (C$^2$) ppm. $^{31}$P NMR (400 MHz, C$_6$D$_6$): $\delta$ = 58.8 ppm.

**Synthesis of 2c:** (C$_2$H$_5$)$_2$PCl (12.2 µL, 0.1 mmol) was added to a solution of **1** (0.082 g, 0.1 mmol) in toluene (10 mL). A colour change from purple to brown was observed. The solution was stirred for two hours at room temperature. Then, the solvent was evaporated in vacuum and the product was extracted with hexane (20 mL). The filtrate was kept at -30°C to afford red-brown crystals. Yield 0.033 g (39 %). C$_{54}$H$_{68}$Cr$_2$N$_4$PCl (942.37): calcd. C 68.74, H 7.26, N 5.94; found C 68.81, H 7.84, N 5.11. $^1$H NMR (250 MHz, C$_6$D$_6$): $\delta$ = 0.88 (t, 6H, $^3J_{HH}$ = 6.8 Hz, H$^{\underline{CH3}CH2P}$), 1.06-1.15 (m, 18H, H$^{CH3\underline{CH2}P}$, H$^{22/23/25/26}$), 1.26 (d, 6H, $^3J_{HH}$ = 6.8 Hz, H$^{22/23/25/26}$), 1.43 (d, 6H, $^3J_{HH}$ = 6.8 Hz, H$^{22/23/25/26}$), 2.07 (s, 6H, H$^{13/14}$), 2.28 (s, 6H, H$^{13/14}$), 3.69 (sep, 2H, $^3J_{HH}$ = 6.5 Hz, H$^{21/24}$), 4.20 (sep, 2H, $^3J_{HH}$ = 6.5 Hz, H$^{21/24}$), 5.78 (d, 2H, $^3J_{HH}$ = 6.5 Hz, H$^3$), 6.26 (tr, 4H, $^3J_{HH}$ = 8.7 Hz, H$^{10/18}$), 6.50 (d, 2H, $^3J_{HH}$ = 6.8 Hz, H$^5$), 6.81 (t, 2H, $^3J_{HH}$ = 6.8 Hz, H$^4$), 7.02-7.16

(m, 8H, H$^{9/11,17/19}$) ppm. $^{13}$C NMR (100 MHz, C$_6$D$_6$): $\delta$ = 13.0 (C$^{\underline{CH3}CH2P}$), 13.8 (C$^{CH3\underline{CH2}P}$), 20.7 (C$^{13/14}$), 21.6 (C$^{13/14}$), 22.9 (C$^{22/23/25/26}$), 25.7 (C$^{22/23/25/26}$), 26.3 (C$^{22/23/25/26}$), 27.6 (C$^{22/23/25/26}$), 28.9 (C$^{21/24}$), 107.2 (C$^3$), 111.2 (C$^5$), 123.5 (C$^{9/11}$), 126.1 (C$^{17/19}$), 127.5 (C$^{18}$), 128.2 (C$^{10}$), 134.3 (C$^{8/12}$), 135.5 (C$^{8/12}$), 136.7 (C$^4$), 145.7 (C$^7$), 147.8 (C$^{16/20}$), 157.2 (C$^6$), 170.0 (C$^2$) ppm. $^{31}$P NMR (400 MHz, C$_6$D$_6$): $\delta$ = 70.8 ppm.

## Acknowledgments

Financial support from the Deutsche Forschungsgemeinschaft (DFG KE 756/20-1) is gratefully acknowledged. E. S. T. is grateful to the Deutsche Akademische Austauschdienst (DAAD) for a Ph.D. scholarship. We thank Dr. Elena Klimkina for her support on $^{31}$P NMR measurements.

## 6.5 References

[1] a) F. Wagner, A. Noor, R. Kempe, *Nat. Chem.* **2009**, *1*, 529-536; b) F. A. Cotton, L. A. Murillo, R. A. Walton, Multiple Bonds Between Metal Atoms, 3rd ed., Springer, Berlin, **2005**.

[2] a) E. P. Kündig, M. Moskovits, G. A. Ozin, *Nature* **1975**, *254*, 503-504; b) W. Klotzbücher, G. A. Ozin, *Inorg. Chem.* **1977**, *16*, 984-987; c) Y. M. Efremov, A. N. Samoilova, L. V. Gurvich, *Opt. Spektrosc.* **1974**, *36*, 654-657; d) V. E. Bondybey, J. H. English, *Chem. Phys. Lett.* **1983**, *94*, 443-447.

[3] a) T. Nguyen, A. D. Sutton, M. Brynda, J. C. Fettinger, G. J. Long, P. P. Power, *Science* **2005**, *310*, 844-847; b) R. Wolf, C. Ni, T. Nguyen, M. Brynda, G. J. Long, A. D. Sutton, R. C. Fischer, J. C. Fettinger, M. Hellman, L. Pu, P. P. Power, *Inorg. Chem.* **2007**, *46*, 11277-11290; c) E. Rivard, P. P. Power, *Inorg. Chem.* **2008**, *46*, 10047-10064.

[4] K. A. Kreisel, G. P. A. Yap, O. Dmitrenko, C. R. Landis, K. H. Theopold, *J. Am. Chem. Soc.* **2007**, *129*, 14162-14163.

[5] a) Y.-C. Tsai, C.-W. Hsu, J.-S. K. Yu, G.-H. Lee, Y. Wang, T.-S. Kuo, *Angew. Chem.* **2008**, *120*, 7250-7253; *Angew. Chem. Int. Ed.* **2008**, *47*, 7250-7253; b) C.-W. Hsu, J.-S. K. Yu, C.-H. Yen, G.-H. Lee, Y. Wang, Y.-C. Tsai, *Angew. Chem.* **2008**, *120*, 10081-10084; *Angew. Chem. Int. Ed.* **2008**, *47*, 9933-9936; c) L.-C. Wu, C.-W. Hsu, Y.-C. Chuang, G.-H. Lee, Y.-C. Tsai, Y. Wang, *J. Phys. Chem. A* **2011**, *115*, 12602–12615; d) Y.-L Huang, D.-Y. Lu, H.-S. Yu, J.-S. K. Yu, C.-W. Hsu, T.-S. Kuo, G.-H. Lee, Y. Wang, Y.-C. Tsai, *Angew. Chem.* **2012**, *124*, 7901-7905; *Angew. Chem. Int. Ed.* **2012**, *51*, 7781-7785.

[6] a) A. Noor, F. R. Wagner, R. Kempe, *Angew. Chem.* **2008**, *120*, 7356-7359; *Angew. Chem. Int. Ed.* **2008**, *47*, 7246-7249; b) A. Noor, G. Glatz, R. Müller, M. Kaupp, S. Demeshko, R. Kempe, *Z. Anorg. Allg. Chem.* **2009**, *635,* 1149-1152; c) A. Noor, R. Kempe, *Chem. Rec.* **2010**, *10,* 413-416.

[7] a) Y.-C. Tsai, H.-Z. Chen, C.-C. Chang, J.-S. K. Yu, G.-H. Lee, Y. Wang, T.-S. Kuo, *J. Am. Chem. Soc.* **2009**, *131*, 12534-12535; b) S.- C. Liu, W.-L. Ke, J.-S. K. Yu, T.-S. Kuo, Y.-C. Tsai, *Angew. Chem.* **2012**, *124*, 6500-6503; *Angew. Chem. Int. Ed.* **2012**, *51*, 6394-6397.

[8] a) A. Noor, G. Glatz, R. Müller, M. Kaupp, S. Demeshko, R. Kempe, *Nat. Chem.* **2009**, *1*, 322-325; b) C. Ni, B. D. Ellis, G. J. Long, P. P. Power, *Chem. Commun.* **2009**, 2332–2334; c) A. Noor, E. Sobgwi Tamne, S. Qayyum, T. Bauer, R. Kempe, *Chem. Eur. J.* **2011**, *17*, 6900-6903; d) J. Shen, G. P. A. Yap, J.-P. Werner, K. H. Theopold, *Chem. Commun.* **2011**, *47*, 12191-12193; e) H.-Z Chen, S.-C. Liu, C.-H. Yen, J.-S. K. Yu, Y.-J. Shieh, T.-S. Kuo, Y.-C. Tsai, *Angew. Chem.* **2012**, *124*, 10488-10492; *Angew. Chem. Int. Ed.* **2012**, *51*, 10342-10346; f) C. Schwarzmaier, A. Noor, G. Glatz, M. Zabel, A. Y. Timoshkin, B. M. Cossairt, C. C. Cummins, R. Kempe, M. Scheer, *Angew. Chem.* **2011**, *123*, 7421-7424; *Angew. Chem. Int. Ed.* **2011**, *50*, 7283-7286.

[9] P. Wei, D. W. Stephan, *Organometallics* **2003**, *22*, 1712-1717.

[10] S. Deeken, G. Motz, R. Kempe, *Z. Anorg. Allg. Chem.* **2007**, *633*, 320-325.

[11] A. Altomare, M. C. Burla, M. Camalli, G. L. Cascarano, C. Giacovazzo, A. Guagliardi, A. G. G. Polidori, R. Spagna, *J. Appl. Cryst.* **1999**, *32*, 115-119.

[12] G. M. Sheldrick, *Acta Cryst. A* **2008**, *64*, 112-122.

[13] L. J. Farrugia, *J. Appl. Cryst.* **1999**, *32*, 837-838.

# 7. List of Publications

The following publications have been published, or submitted or are to be submitted during the work of this thesis:

1- Awal Noor, <u>Emmanuel Sobgwi Tamne</u>, Sadaf Qayyum, Tobias Bauer, Rhett Kempe, *Chem. Eur. J.* **2011**, *17*, 6900-6903.
"Cycloaddition Reactions of a Cr-Cr Quintuple Bond"

2- <u>Emmanuel Sobgwi Tamne</u>, Awal Noor, Sadaf Qayyum, Tobias Bauer, Rhett Kempe, *Inorg. Chem.*, accepted for publication.
"Quintuple Bond Reactivity towards Group 16 and 17 Elements: Addition vs. Insertion"

3- Awal Noor, <u>Emmanuel Sobgwi Tamne</u>, Benjamin Oelkers, Tobias Bauer, Serhiy Demeshko, Franc Meyer, Frank Heinemann, Rhett Kempe, Submitted to *Angew. Chem. Int. Ed.*
"Von Chrom-Chrom-Fünffachbindungen über molekulare Quadrate zu porösen Materialien"

4- <u>Emmanuel Sobgwi Tamne</u>, Awal Noor, Tobias Bauer, Rhett Kempe, to be submitted in *Z. Anorg. Allg. Chem.*
"Reactions of a Cr-Cr Quintuple Bond with Phosphine Ligands"

## 8. Acknowledgements

I want to thank God for without Him I would not have been able to achieve this Ph.D.

I would like to express my gratitude to my academic supervisor Prof. Dr. Rhett Kempe for accepting me as his Ph.D student and for giving me the opportunity to work on a very interesting subject, the excellent working conditions and the great scientific independence he has granted me.

I would also like to sincerely thank the Deutscher Akademischer AustauschDienst (**DAAD**) for offering me a scholarship without which I certainly would not have studied in such good conditions.

It is time to express my immense gratitude to Dr. Awal Noor, who introduced me to the world of Schlenk technique. I thank him also for his patience, guidance and suggestions during my work. Without all these efforts from him, accomplishment of this work would have been extremely difficult.

I would like to thank Dr. Germund Glatz, Tobias Bauer and Isabelle Haas for their help in determining the molecular structures of several new compounds and Dr. Benoît Blank for his patience and help for the integration in lab and in the city during my first days.

Thank you to Dr. Elena Klimkina for her multiple helps whenever needed; to Dr. Benjamin Oelkers for the correction of the manuscripts and, to Justus Hermannsdörfer for his help in translating the summary.

I am particularly grateful to my lab-mates Dr. Sadaf Qayyum and Prashant Kumar for the many interesting scientific discussions and advices.

Thanks are also to Walter Kremnitz, Marlies Schilling, Anna Maria Dietel and Heidi Maisel, who did the administrative work, provided dry solvents and maintained the analytic equipments.

I would like to thank my colleagues: Dr. Christine Denner, Dr. Torsten Irrgang, Dr. Winfried. P. Kretschmer, Adam Sobacynski, Johannes Obenauf, Stefan Michlik, Saravana Pillai, Theresa Winkler, Susanne Ruch, Sina Rösler, Muhammad Zaheer, and Muhammad Hafeez for their cooperation and nice company. Also it's the time to thank all the people who have contributed to my work in one way or the other.

My heartiest thanks to my beloved wife Elise Chimene Matoukam for her absolute confidence in me and for always being with me.

I express my acknowledgements to my family for their never ending support. Particularly to my father Charles Tamne, my sisters Gisele Flora, Sabine Laure, Lucie Claire and Nadege Stephanie. Thank you to Mr and Mrs Noumsi, Kengne, Wambo, and Kuate for their constant support.
My in-laws especially my mother-in-law are warmly thanked for their sincere wishes and prayers.

Finally, I would like to express my gratitude to Dr. Harold Tanh Jeazet and Stephan Raoul Noumen for their constant support and help.

Thank you to all my friends Astride Kemmoe, Paul Djomgoue, Dr. Laure Peem, Dr. Zoila Epossi, Dr. Eyane Meva'a, Dr. Eric Anchimbe, and Gilbert Ndi Shang for the great moments we had together.

# I want morebooks!

Buy your books fast and straightforward online - at one of the world's fastest growing online book stores! Environmentally sound due to Print-on-Demand technologies.

Buy your books online at

**www.get-morebooks.com**

Kaufen Sie Ihre Bücher schnell und unkompliziert online – auf einer der am schnellsten wachsenden Buchhandelsplattformen weltweit! Dank Print-On-Demand umwelt- und ressourcenschonend produziert.

Bücher schneller online kaufen

**www.morebooks.de**

VDM Verlagsservicegesellschaft mbH
Heinrich-Böcking-Str. 6-8
D - 66121 Saarbrücken

Telefax: +49 681 93 81 567-9

info@vdm-vsg.de
www.vdm-vsg.de

Printed by Books on Demand GmbH, Norderstedt / Germany